Microstructural Design and Processing Control of Advanced Ceramics

Microstructural Design and Processing Control of Advanced Ceramics

Editors

Qingyuan Wang
Yu Chen

MDPI • Basel • Beijing • Wuhan • Barcelona • Belgrade • Manchester • Tokyo • Cluj • Tianjin

Editors
Qingyuan Wang
College of Architecture and
Environment
Sichuan University
Chengdu
China

Yu Chen
School of Mechanical
Engineering
Chengdu University
Chengdu
China

Editorial Office
MDPI
St. Alban-Anlage 66
4052 Basel, Switzerland

This is a reprint of articles from the Special Issue published online in the open access journal *Materials* (ISSN 1996-1944) (available at: www.mdpi.com/journal/materials/special_issues/Microstructural_Processing_Ceramics).

For citation purposes, cite each article independently as indicated on the article page online and as indicated below:

LastName, A.A.; LastName, B.B.; LastName, C.C. Article Title. *Journal Name* **Year**, *Volume Number*, Page Range.

ISBN 978-3-0365-6625-2 (Hbk)
ISBN 978-3-0365-6624-5 (PDF)

© 2023 by the authors. Articles in this book are Open Access and distributed under the Creative Commons Attribution (CC BY) license, which allows users to download, copy and build upon published articles, as long as the author and publisher are properly credited, which ensures maximum dissemination and a wider impact of our publications.

The book as a whole is distributed by MDPI under the terms and conditions of the Creative Commons license CC BY-NC-ND.

Contents

About the Editors vii

Preface to "Microstructural Design and Processing Control of Advanced Ceramics" ix

Yu Chen and Qingyuan Wang
Microstructural Design and Processing Control of Advanced Ceramics
Reprinted from: *Materials* 2023, 16, 905, doi:10.3390/ma16030905 1

Huajiang Zhou, Shaozhao Wang, Daowen Wu, Qiang Chen and Yu Chen
Microstructures and Electrical Conduction Behaviors of Gd/Cr Codoped Bi_3TiNbO_9 Aurivillius Phase Ceramic
Reprinted from: *Materials* 2021, 14, 5598, doi:10.3390/ma14195598 7

Shaozhao Wang, Huajiang Zhou, Daowen Wu, Lang Li and Yu Chen
Effects of Oxide Additives on the Phase Structures and Electrical Properties of $SrBi_4Ti_4O_{15}$ High-Temperature Piezoelectric Ceramics
Reprinted from: *Materials* 2021, 14, 6227, doi:10.3390/ma14206227 23

Daowen Wu, Huajiang Zhou, Lingfeng Li and Yu Chen
Gd/Mn Co-Doped $CaBi_4Ti_4O_{15}$ Aurivillius-Phase Ceramics: Structures, Electrical Conduction and Dielectric Relaxation Behaviors
Reprinted from: *Materials* 2022, 15, 5810, doi:10.3390/ma15175810 35

Rui Zhang, Biao Chen, Fuyan Liu, Miao Sun, Huiming Zhang and Chenlong Wu
Microstructure and Mechanical Properties of Composites Obtained by Spark Plasma Sintering of Ti_3SiC_2-15 vol.%Cu Mixtures
Reprinted from: *Materials* 2022, 15, 2515, doi:10.3390/ma15072515 47

Rui Zhang, Huiming Zhang and Fuyan Liu
Microstructure and Tribological Properties of Spark-Plasma-Sintered Ti_3SiC_2-Pb-Ag Composites at Elevated Temperatures
Reprinted from: *Materials* 2022, 15, 1437, doi:10.3390/ma15041437 57

Kunyang Fan, Wenhuang Jiang, Jesús Ruiz-Hervias, Carmen Baudín, Wei Feng and Haibin Zhou et al.
Effect of Al_2TiO_5 Content and Sintering Temperature on the Microstructure and Residual Stress of Al_2O_3–Al_2TiO_5 Ceramic Composites
Reprinted from: *Materials* 2021, 14, 7624, doi:10.3390/ma14247624 69

Chen Ye and Yu Huan
Studies on Electron Escape Condition in Semiconductor Nanomaterials via Photodeposition Reaction
Reprinted from: *Materials* 2022, 15, 2116, doi:10.3390/ma15062116 93

Yang Li, Min Ge, Shouquan Yu, Huifeng Zhang, Chuanbing Huang and Weijia Kong et al.
Characterization and Microstructural Evolution of Continuous BN Ceramic Fibers Containing Amorphous Silicon Nitride
Reprinted from: *Materials* 2021, 14, 6194, doi:10.3390/ma14206194 105

Yaming Zhang, Bingbing Li and Yanmin Jia
High Humidity Response of Sol–Gel-Synthesized $BiFeO_3$ Ferroelectric Film
Reprinted from: *Materials* 2022, 15, 2932, doi:10.3390/ma15082932 117

Yubai Ma, Mei Li and Fangqiu Zu
The Tribological Behaviors in Zr-Based Bulk Metallic Glass with High Heterogeneous Microstructure
Reprinted from: *Materials* **2022**, *15*, 7772, doi:10.3390/ma15217772 . . . . **127**

About the Editors

Qingyuan Wang

Prof. Wang Qingyuan is a full Professor of Sichuan University since 2001. He is currently the president of Chengdu University. He received his PhD in 1998 from Ecole Centrale Paris (University Paris-Saclay), France, followed by postdoctoral experience from 1999 to 2003 in the Faculty of Engineering of Purdue University, and JSPS fellow at Kagoshima University, Japan. He served as a visiting professor at Univ Paris Sud, Clausthal TU, Univ New Hampshire and Honorary Professor of Univ Liverpool, RMIT. He is also a fellow of the International Association of Advanced Materials (IAAM).

His research activity is focused on Materials, Mechanics and Structures, Super Long Life Durability, Low carbon technologies and Sustainable development. He has advised more than 10 Postdoc, 30 PhD and 40 Master students. He is chairman of organizing committee for the 6th Inter Conf on Very High Cycle fatigue and the 9th Chinese Congress on Mechanics. Prof. Wang has published over 300 papers in SCI index journals, of which more than 100 articles published on Top Journals such as Inter J Fatigue, Mater Design, Acta Mater, Small, Matter, Nature Commun, etc. He is one of the Most Cited Chinese Researchers (Elsevier 2014-2021) and among the World's Top 2% Scientists 2020-2022 and 1960-2021 (Mendeley Data).

Yu Chen

Dr. Yu Chen achieved both a Bachelor's degree (2006) and Master's degree (2009) in Materials Science at Sichuan University, and achieved a Ph.D. degree (2016) in Solid Mechanics after a joint educational project between Sichuan University and Liverpool University. He is currently an Associate Professor in Chengdu University. His main research interests focus on dielectric/piezoelectric/ferroelectric ceramics and their applications in smart sensing/actuating devices and energy harvest/storage/transformation systems.

To date, he has published over 40 papers in journals including *Materials & Design*, *Acta Materialia*, *Scripta Materialia*, the *Journal of Advanced Ceramics*, the *Journal of Materiomics*, the *Journal of the American Ceramic Society*, etc. Additionally, he has also obtained eight Chinese patents. Now, he acts as a Topical Advisory Panel Member of *Materials*, an Associated Editorial Committee Member for the *Journal of Advanced Ceramics*, as well as a Youth Member of the Editorial Board for the *Journal of Advanced Dielectrics*. In 2014, he won the Outstanding Paper Award from the Joint Conference of 9th Asian Meeting on Ferroelectrics & 9th Asian Meeting on Electroceramics (AMF-AMEC-2014). In 2018, he received the Innovation Talent Award of Ceramic Technology from The Chinese Ceramic Society. In 2022, his research group's project (High-temperature and High-performance Piezoelectric Ceramics) won the National Bronze Award in the 8th China International College Students' "Internet+" Innovation and Entrepreneurship Competition.

Preface to "Microstructural Design and Processing Control of Advanced Ceramics"

The global demand for ceramic materials with wide-ranging applications in the environment, precision tools, biomedical, electronics, and environmental fields is on the rise. Advanced ceramics can be elaborated as processed ceramic materials with task-specific properties such as high mechanical performances and/or unique functional properties. In view of their various applications covering thermal conductors, cutting tools, automotive/atomic energy/electronic/biomedical devices, micro-electro-mechanical systems, environmental and aerospace engineering, advanced ceramics represent an important technology that has a considerable impact for a large variety of industries, branches and markets. It is considered as an enabling technology that has the potential to deliver high-value contributions for solving the challenges of our future.

Advanced ceramics can also be referred to as high performance ceramics, high-tech ceramics, engineering ceramics, fine ceramics, and technical ceramics. Such materials may be oxides, carbides, nitrides, silicides, borides, etc. They are basically crystalline materials of rigorously controlled composition, microstructure and manufactured with detailed regulation from highly refined or/and characterized raw materials that give precisely specified attributes. The solid-state reactions are the most widely used processes for the mass-production of cost-efficient ceramic powders. Additionally, the sol-gel technique, CVD, and polymer pyrolysis have been applied to generate high-purity ceramics precursor with defined properties. In general, the chemical compositions, microstructural characteristics, fabrication methods, and processing conditions of advanced ceramics impel their characteristics including corrosion-resistant, outstanding optical and electrical properties, hardness and anti-aging.

This reprint contains one editorial paper and ten research-type papers that reported the results of several researches regarding functional ceramics, structural ceramics, ceramic composites, ceramic coatings and special glasses. The complex chemical compositions, novel and controllable fabrication process, subtle and designable meso/micro/nano-structures, unique physical and chemical properties as well as potential applications of these materials will be presented to the readers.

Qingyuan Wang and Yu Chen
Editors

Editorial
Microstructural Design and Processing Control of Advanced Ceramics

Yu Chen [1,*] and Qingyuan Wang [1,2]

1 School of Mechanical Engineering, Chengdu University, Chengdu 610106, China
2 College of Architecture and Environment, Sichuan University, Chengdu 610065, China
* Correspondence: chenyu01@cdu.edu.cn

Advanced ceramics are referred to in various parts of the world as technical ceramics, high-tech ceramics, and high-performance ceramics. They represent an important technology that has considerable impacts for a large variety of industries, branches and markets. It is considered as an enabling technology that has the potential to deliver high-value contributions for solving the challenges of our future. From a general point of view, the advanced ceramics sector comprises the following categories [1]:

(1) Functional ceramics: Electrical and magnetic ceramics (i.e., dielectrics, piezoelectrics, ferromagnetics), ionic conductors and superconductive ceramics.
(2) Structural ceramics: Monoliths and composites, e.g., oxides, nitrides, carbides, borides, and composite materials based on these materials.
(3) Bioceramics: e.g., hydroxyapatite and alumina.
(4) Ceramic coatings: Oxides, nitrides, carbides, borides, cermets and diamond-like coatings, deposited by technologies such as spraying, vapor deposition and sol-gel coating.
(5) Special glasses: Processed flat glass, fire resistant glazing and glasses for optoelectronics.

Figure 1 shows some representative advanced ceramics developed in recent years.

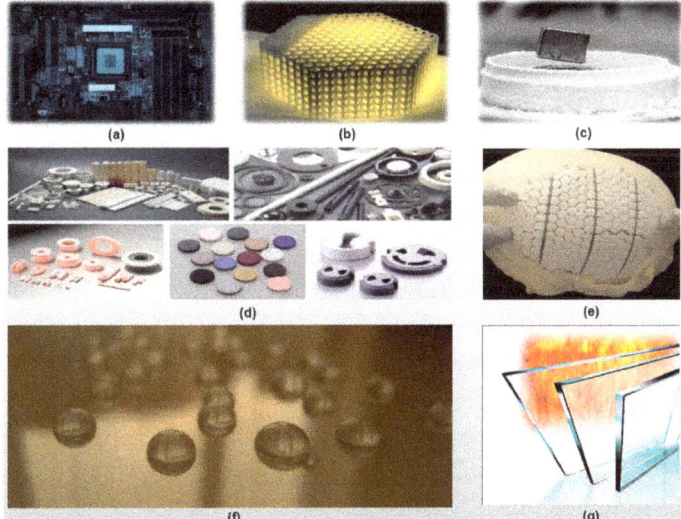

Figure 1. Some representative advanced ceramics developed in recent years: (**a**) Electronic ceramics, (**b**) 3-D printed high-strength ceramics, (**c**) Superconductive ceramics, (**d**) Various structural ceramics, (**e**) HAP bioceramics, (**f**) Nano-ceramic coatings, (**g**) Fire-resistant glasses.

Advanced ceramics are usually composed of dense and fine-grained microstructures, thus can also be called fine ceramics. Due to their special mechanical performances and/or unique functional properties, advanced ceramics are useful in various applications covering thermal conductors, cutting tools, auto-motive/atomic energy/electronic/biomedical devices, energy conversion/sensor/actuation systems, and environmental and aerospace engineering. Japan Fine Ceramics Association (JFCA) published the "FC Roadmap 2050 (2021 edition)" in December 2021 [2]. The world market of the production of the fine ceramics industry is about USD 70 billion in 2018. Although the global production of automobiles and smartphones fell far short of the previous year's level due to the impact of COVID-19 in recent years, fine ceramics output increased due to higher demand for semiconductors used in equipment, such as PCs and memory devices [3]. It is well known that after the outbreak of COVID-19 in 2020, ceramic materials have played an important role in combating the epidemic, especially piezoelectric ceramics, which play an important role in respirators, masks and advanced ultrasonic medical equipment (Figure 2). Besides the applications in the medicine field, piezoelectric ceramics can be also used as sensitive materials for various kinds of electroacoustic and piezoelectric devices, including sensors, detonators, micro-displacement actuators, ultrasonic transducers etc. In 2021, Global Industry Analysts, Inc published the latest global sales market forecasts of advanced ceramics. In the global sales market, China and the United States have large shares. An average annual growth of 6.3% is expected from 2020 to 2027 [4].

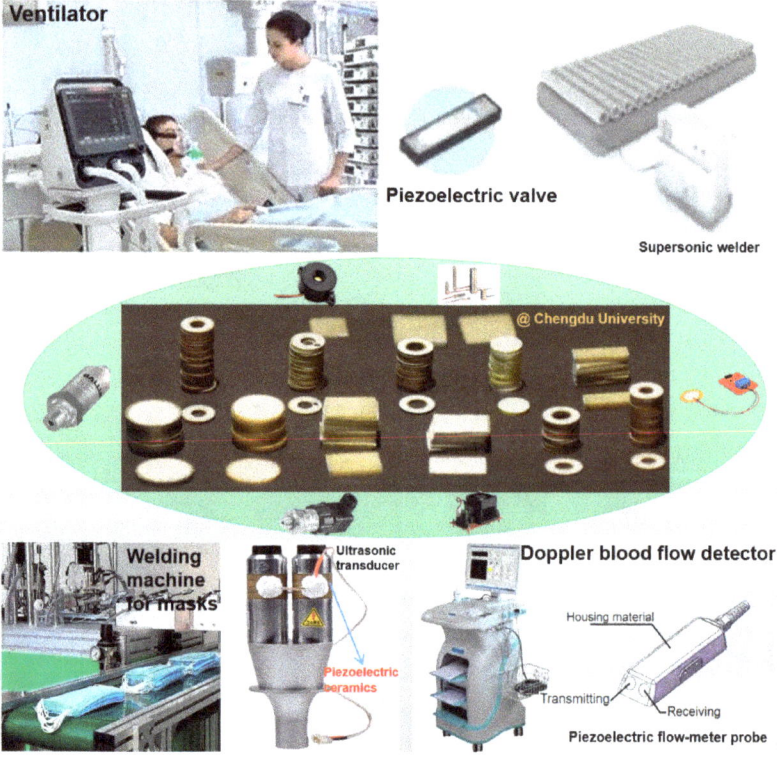

Figure 2. Applications of piezoelectric ceramics in the medicine field.

Advanced ceramics possess the tunable compositions and designable microstructures. First, advanced ceramics tend to lack a glassy component; i.e., they are "basically crystalline". Second, microstructures are usually highly engineered, meaning that grain sizes, grain shapes, porosity, and phase distributions (for instance, the arrangements of second phases such as whiskers and fibers) can be carefully planned and controlled. Such planning and control require "detailed regulation" of composition and processing. Finally, such advanced ceramics with both well-designed microstructures tend to exhibit unique or superior functional attributes that can be "precisely specified" by careful processing and quality control [5]. There are many examples regarding the unique electrical properties such as excellent piezoelectricity, superconductivity or superior mechanical performances, including enhanced toughness or high-temperature strength, which were achieved by the microstructural design for ferroelectric/piezoelectric ceramics, bioceramics, structural ceramics, metal–ceramic composites, etc. [6–12].

The preparation of advanced ceramic components involves the heating process of ceramic powders, which must undergo special handling to control the heterogeneity, chemical compositions, purity, particle size, particle size distribution (PSD) and specific shape [13]. The aforementioned factors play a significant role in the properties of finished ceramic components. Moreover, the preparation of advanced ceramics usually involves more sophisticated processing steps. In short, processes become rather complex and can differ for various applications during the development of advanced ceramics [14]. The fabrication methods and processing conditions of advanced ceramics also impel their characteristics including excellent thermal properties, optical and electrical properties, corrosion-resistant, mechanical strength, hardness and anti-aging [15–17].

For example, in the editors' laboratory, the PMN-29PT-1.6Gd ferroelectric ceramics were fabricated by a A-site modified oxide precursor method with two processing steps [18]. The removal of polar defect pairs with Gd-doping was considered to promote greater field-induced PNR reorientation and thereby increase the permittivity greatly. The combination of exquisite microstructural design and optimum processing control helped the ceramics to achieve an ultra-high piezoelectric coefficient (d_{33}) up to 1210 pC/N, associated with a high dielectric permittivity of ε_r = 7059 at room temperature. Figure 3 shows the synthetic strategy of the PMN-29PT-1.6Gd ceramics designed by the authors.

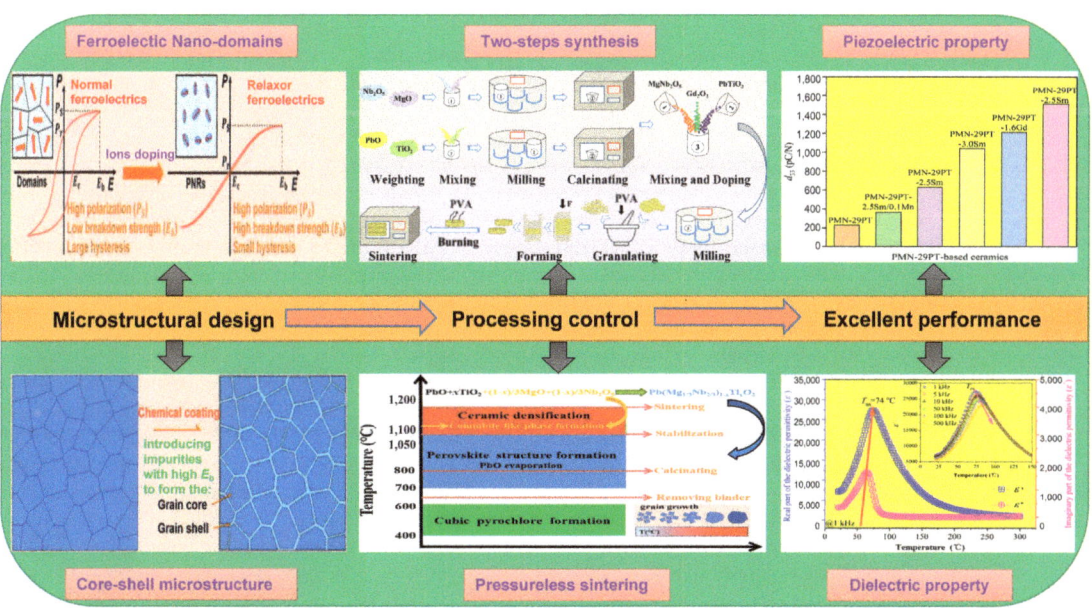

Figure 3. Synthetic strategy of the PMN-29PT-1.6Gd ferroelectric ceramics with excellent performances.

Because of the attention to microstructural design and processing control, advanced ceramics are often high value-added products. Developments in advanced ceramic processing continue at a rapid pace, constituting what can be considered a revolution in the kind of materials and properties obtained.

This special issue contains ten papers that reported the results of several studies on functional ceramics, structural ceramics, ceramic composites, ceramic coatings and special glasses. The exquisite chemical compositions, novel and controllable fabrication process, subtle and designable meso/micro/nano-structures, unique physical and chemical properties as well as potential applications of these materials will be presented to the readers.

The editors' research group focused on the preparation, microstructures and electrical behaviors of the modified bismuth layer ferroelectric (BLSFs) ceramics. The authors of [19] studied the microstructures and electrical conduction behaviors of Gd/Cr co-doped Bi_3TiNbO_9 Aurivillius-phase ceramics. The authors of [20] reported the effects of oxide additives on the phase structures and electrical properties of $SrBi_4Ti_4O_{15}$ high-temperature piezoelectric ceramics. The authors of [21] revealed the structures, electrical conduction and dielectric relaxation behaviors of the Gd/Mn co-doped $CaBi_4Ti_4O_{15}$ Aurivillius-phase ceramics. By means of the chemical (substituting ions and oxide additives) modification and microstructural (phase composition and grain size/orientation) design for these three kinds of BLSFs ceramics, the electrical properties of materials were greatly improved. Such BLSFs ceramics are expected to obtain wide applications in the piezoelectric sensors with an operating temperature exceeding 500 °C.

The authors of [22] reported the microstructure and mechanical properties of composites obtained by spark plasma sintering of Ti_3SiC_2-15 vol.%Cu mixtures, and the microstructure and tribological properties of spark-plasma sintered Ti_3SiC_2-Pb-Ag composites at elevated temperatures were further investigated in [23]. The authors of [24] explored the effect of Al_2TiO_5 content and sintering temperature on the microstructure and residual stress of Al_2O_3-Al_2TiO_5 ceramic composites, and the effects of material design and sintering process on residual stresses were expounded from macro- and micro-levels. The authors of [25] studied the electron escape condition in semiconductor nanomaterials via photodeposition reaction. The authors of [26] prepared the boron nitride ceramic fibers containing amounts of silicon nitride using hybrid precursors of PBN and PCS via melt-spinning, curing, decarburization under NH_3 to 1000 °C and pyrolysis up to 1600 °C under N_2. The authors of [27], via a facile sol-gel method, explored the effects of the relative humidity (RH) on the $BiFeO_3$ film in terms of capacitance, impedance and current-voltage (I–V). The authors of [28] investigated the tribological behaviors in Zr-based bulk metallic glass with a high heterogeneous microstructure.

In the manufacturing process of advanced ceramics, such as electronic ones, which have a high global market share, complicated physical and chemical changes occur. The characteristics of advanced ceramics, such as performance, reliability, and durability, are determined by the microstructure resulting from the transformation of the manufacturing process.

The experimental research and theoretic analysis aiming at some common scientific and technological problems in advanced ceramics presented above constitute the thematic scope of this Special Issue, entitled "Microstructural Design and Processing Control of Advanced Ceramics". In conclusion, the published papers demonstrate the microstructures and processes' relevance of the topics dealt with. The main purpose of this Special Issue is to publish significant papers presenting advanced research in the field of ceramic materials and ceramic composites with excellent functional and/or mechanical properties and broad applications.

Acknowledgments: All contributing authors and the editorial team of *Materials* are acknowledged for their continued support for this Special Issue.

Conflicts of Interest: The author declares no conflict of interest.

References

1. Rödel, J.; Kounga, A.B.; Weissenberger-Eibl, M.; Koch, D.; Bierwisch, A.; Rossner, W.; Hoffmann, M.J.; Danzer, R.; Schneider, G. Development of a roadmap for advanced ceramics: 2010–2025. *J. Eur. Ceram. Soc.* **2009**, *29*, 1549–1560. [CrossRef]
2. Japan Fine Ceramics Association. *FC Roadmap 2050 2021 Edition*; Japan Fine Ceramics Association: Minato-ku, Tokyo, 2021; pp. 1–62.
3. Takemura, H.; Fukushima, H. Recent trends of advanced ceramics industry and Fine Ceramics Roadmap 2050. *Int. J. Appl. Ceram. Technol.* **2022**, 1–8. [CrossRef]
4. Global Industry Analysts, Inc. *Advanced Ceramics Global Market Trajectory & Analytics August*; Global Industry Analysts, Inc.: San Jose, CA, USA, 2021; p. II-74.
5. Saini, P.K.; Malhi, G.S.; Gupta, A. Study the Influence of EDM Parameters on Cobalt-Bonded Tungsten Carbide Conductive Ceramic. *Int. J. Eng. Manag. Res.* **2015**, *5*, 69–72.
6. Chen, Y.; Xie, S.; Wang, Q.; Fu, L.; Nie, R.; Zhu, J. Correlation between microstructural evolutions and electrical/mechanical behaviors in Nb/Ce co-doped Pb($Zr_{0.52}Ti_{0.48}$)O_3 ceramics at different sintering temperatures. *Mater. Res. Bull.* **2017**, *94*, 174–182. [CrossRef]
7. Chen, Y.; Zhou, H.; Wang, Q.; Zhu, J. Doping level effects in Gd/Cr co-doped Bi_3TiNbO_9 Aurivillius-type ceramics with improved electrical properties. *J. Mater.* **2022**, *8*, 906–917. [CrossRef]
8. Tan, Z.; Xie, S.; Jiang, L.; Xing, J.; Chen, Y.; Zhu, J.; Xiao, D.; Wang, Q. Oxygen octahedron tilting, electrical properties and mechanical behaviors in alkali niobate-based lead-free piezoelectric ceramics. *J. Materiomics* **2019**, *5*, 372–384. [CrossRef]
9. Yang, D.; Zhou, Y.; Yan, X.; Wang, H.; Zhou, X. Highly conductive wear resistant Cu/Ti_3SiC_2(TiC/SiC) co-continuous composites via vacuum infiltration process. *J. Adv. Ceram.* **2020**, *9*, 83–93. [CrossRef]
10. Feng, C.; Zhang, K.; He, R.; Ding, G.; Xia, M.; Jin, X.; Xie, C. Additive manufacturing of hydroxyapatite bioceramic scaffolds: Dispersion, digital light processing, sintering, mechanical properties, and biocompatibility. *J. Adv. Ceram.* **2020**, *9*, 360–373. [CrossRef]
11. Zhang, R.; Pei, J.; Han, Z.J.; Wu, Y.; Zhao, Z.; Zhang, B.P. Optimal performance of $Cu_{1.8}S_{1-x}Te_x$ thermoelectric materials fabricated via high-pressure process at room temperature. *J. Adv. Ceram.* **2020**, *9*, 535–543. [CrossRef]
12. Nie, G.; Li, Y.; Sheng, P.; Zuo, F.; Wu, H.; Liu, L.; Deng, X.; Bao, Y.; Wu, S. Microstructure refinement-homogenization and flexural strength improvement of Al_2O_3 ceramics fabricated by DLP-stereolithography integrated with chemical precipitation coating process. *J. Adv. Ceram.* **2021**, *10*, 790–808. [CrossRef]
13. Otitoju, T.A.; Okoye, P.U.; Chen, G.; Li, Y.; Okoye, M.O.; Li, S. Advanced ceramic components: Materials, fabrication, and applications. *J. Ind. Eng. Chem.* **2020**, *85*, 34–65. [CrossRef]
14. Salamon, D. Advanced ceramics. In *Advanced Ceramics for Dentistry*; Butterworth-Heinemann: Oxford, UK, 2014; pp. 103–122.
15. Lorenz, M.; Travitzky, N.; Rambo, C.R. Effect of processing parameters on in situ screen printing-assisted synthesis and electrical properties of Ti_3SiC_2-based structures. *J. Adv. Ceram.* **2021**, *10*, 129–138. [CrossRef]
16. Veerapandiyan, V.; Benes, F.; Gindel, T.; DeLuca, M. Strategies to Improve the Energy Storage Properties of Perovskite Lead-Free Relaxor Ferroelectrics: A Review. *Materials* **2020**, *13*, 5742. [CrossRef]
17. Heydari, L.; Lietor, P.F.; Corpas-Iglesias, F.A.; Laguna, O.H. Ti(C,N) and WC-Based Cermets: A Review of Synthesis, Properties and Applications in Additive Manufacturing. *Materials* **2021**, *14*, 6786. [CrossRef]
18. You, R.; Zhang, D.; Fu, L.; Chen, Y. Dielectric relaxation and electromechanical degradation of gd-doped pmn-29pt fer-roelectric ceramics with ultrahigh piezoelectricity. *J. Vib. Eng. Technol.* **2022**, *10*, 639–647. [CrossRef]
19. Zhou, H.; Wang, S.; Wu, D.; Chen, Q.; Chen, Y. Microstructures and Electrical Conduction Behaviors of Gd/Cr Codoped Bi_3TiNbO_9 Aurivillius Phase Ceramic. *Materials* **2021**, *14*, 5598. [CrossRef]
20. Wang, S.; Zhou, H.; Wu, D.; Li, L.; Chen, Y. Effects of Oxide Additives on the Phase Structures and Electrical Properties of $SrBi_4Ti_4O_{15}$ High-Temperature Piezoelectric Ceramics. *Materials* **2021**, *14*, 6227. [CrossRef] [PubMed]
21. Wu, D.; Zhou, H.; Li, L.; Chen, Y. Gd/Mn Co-Doped $CaBi_4Ti_4O_{15}$ Aurivillius-Phase Ceramics: Structures, Electrical Conduction and Dielectric Relaxation Behaviors. *Materials* **2022**, *15*, 5810. [CrossRef]
22. Zhang, R.; Chen, B.; Liu, F.; Sun, M.; Zhang, H.; Wu, C. Microstructure and Mechanical Properties of Composites Obtained by Spark Plasma Sintering of Ti_3SiC_2-15 vol.%Cu Mixtures. *Materials* **2022**, *15*, 2515. [CrossRef] [PubMed]
23. Zhang, R.; Zhang, H.; Liu, F. Microstructure and Tribological Properties of Spark-Plasma-Sintered Ti_3SiC_2-Pb-Ag Composites at Elevated Temperatures. *Materials* **2022**, *15*, 1437. [CrossRef]
24. Fan, K.; Jiang, W.; Ruiz-Hervias, J.; Baudín, C.; Feng, W.; Zhou, H.; Bueno, S.; Yao, P. Effect of Al2TiO5 Content and Sintering Temperature on the Microstructure and Residual Stress of Al_2O_3-Al_2TiO_5 Ceramic Composites. *Materials* **2021**, *14*, 7624. [CrossRef] [PubMed]
25. Ye, C.; Huan, Y. Studies on Electron Escape Condition in Semiconductor Nanomaterials via Photodeposition Reaction. *Materials* **2022**, *15*, 2116. [CrossRef]
26. Li, Y.; Ge, M.; Yu, S.; Zhang, H.; Huang, C.; Kong, W.; Wang, Z.; Zhang, W. Characterization and Microstructural Evolution of Continuous BN Ceramic Fibers Containing Amorphous Silicon Nitride. *Materials* **2021**, *14*, 6194. [CrossRef] [PubMed]

27. Zhang, Y.; Li, B.; Jia, Y. High Humidity Response of Sol-Gel-Synthesized $BiFeO_3$ Ferroelectric Film. *Materials* **2022**, *15*, 2932. [CrossRef]
28. Ma, Y.; Li, M.; Zu, F. The Tribological Behaviors in Zr-Based Bulk Metallic Glass with High Heterogeneous Microstructure. *Materials* **2022**, *15*, 7772. [CrossRef] [PubMed]

Disclaimer/Publisher's Note: The statements, opinions and data contained in all publications are solely those of the individual author(s) and contributor(s) and not of MDPI and/or the editor(s). MDPI and/or the editor(s) disclaim responsibility for any injury to people or property resulting from any ideas, methods, instructions or products referred to in the content.

Article

Microstructures and Electrical Conduction Behaviors of Gd/Cr Codoped Bi₃TiNbO₉ Aurivillius Phase Ceramic

Huajiang Zhou [1], Shaozhao Wang [1], Daowen Wu [1], Qiang Chen [2] and Yu Chen [1,2,*]

[1] School of Mechanical Engineering, Chengdu University, Chengdu 610106, China; zhouhj1996@163.com (H.Z.); wangsz0908@163.com (S.W.); wdw132812976202021@163.com (D.W.)
[2] College of Materials Science and Engineering, Sichuan University, Chengdu 610065, China; cqscu@scu.edu.cn
* Correspondence: chenyu01@cdu.edu.cn; Tel.: +86-28-8461-6169

Abstract: In this work, a kind of Gd/Cr codoped Bi_3TiNbO_9 Aurivillius phase ceramic with the formula of $Bi_{2.8}Gd_{0.2}TiNbO_9$ + 0.2 wt% Cr_2O_3 (abbreviated as BGTN−0.2Cr) was prepared by a conventional solid-state reaction route. Microstructures and electrical conduction behaviors of the ceramic were investigated. XRD and SEM detection found that the BGTN−0.2Cr ceramic was crystallized in a pure Bi_3TiNbO_9 phase and composed of plate-like grains. A uniform element distribution involving Bi, Gd, Ti, Nb, Cr, and O was identified in the ceramic by EDS. Because of the frequency dependence of the conductivity between 300 and 650 °C, the electrical conduction mechanisms of the BGTN−0.2Cr ceramic were attributed to the jump of the charge carriers. Based on the correlated barrier hopping (CBH) model, the maximum barrier height W_M, dc conduction activation energy E_c, and hopping conduction activation energy E_p were calculated with values of 0.63 eV, 1.09 eV, and 0.73 eV, respectively. Impedance spectrum analysis revealed that the contribution of grains to the conductance increased with rise in temperature; at high temperatures, the conductance behavior of grains deviated from the Debye relaxation model more than that of grain boundaries. Calculation of electrical modulus further suggested that the degree of interaction between charge carriers β tended to grow larger with rising temperature. In view of the approximate relaxation activation energy (~1 eV) calculated from Z'' and M'' peaks, the dielectric relaxation process of the BGTN−0.2Cr ceramic was suggested to be dominated by the thermally activated motion of oxygen vacancies as defect charge carriers. Finally, a high piezoelectricity of d_{33} = 18 pC/N as well as a high resistivity of $\rho_{dc} = 1.52 \times 10^5$ Ω cm at 600 °C provided the BGTN−0.2Cr ceramic with promising applications in the piezoelectric sensors with operating temperature above 600 °C.

Keywords: Aurivillius phase ceramic; Bi_3TiNbO_9; electrical conduction behaviors; impedance spectrum; electrical modulus

Citation: Zhou, H.; Wang, S.; Wu, D.; Chen, Q.; Chen, Y. Microstructures and Electrical Conduction Behaviors of Gd/Cr Codoped Bi₃TiNbO₉ Aurivillius Phase Ceramic. *Materials* **2021**, *14*, 5598. https://doi.org/10.3390/ma14195598

Academic Editor: Andres Sotelo

Received: 31 August 2021
Accepted: 23 September 2021
Published: 26 September 2021

Publisher's Note: MDPI stays neutral with regard to jurisdictional claims in published maps and institutional affiliations.

Copyright: © 2021 by the authors. Licensee MDPI, Basel, Switzerland. This article is an open access article distributed under the terms and conditions of the Creative Commons Attribution (CC BY) license (https:// creativecommons.org/licenses/by/ 4.0/).

1. Introduction

Bismuth layer—structured ferroelectric (BLSF) compounds (also called Aurivillius phase compounds), with the general formula $[Bi_2O_2][A_{m-1}B_mO_{3m+1}]$, where A with a dodecahedral coordination is a low valence element (less than or equal to trivalent), B with a octahedral coordination is a transition metal element (e.g., Cr^{3+}, Ce^{4+}, Ti^{4+}, Nb^{5+}, Ta^{5+}, W^{6+}), and $1 \leq m \leq 6$, constructed by $m[ABO_3]^{2-}$ layers that alternate with $[Bi_2O_2]^{2+}$ layers [1–3]. Due to their relatively high Curie point (T_C) and excellent fatigue resistance, they have attracted extensive attention for their potential application in high-temperature piezoelectric systems [4,5] and ferroelectric random-access memory (FeRAM) [6]. However, the piezoelectricity of these compounds is limited because of the 2D orientation restriction of their spontaneous polarization (P_s) rotation and their high coercive fields (E_c), and high conductivity also restricts their applications in high temperature environment [7]. The conductivity mechanisms of most ferroelectric materials are approximately divided into three categories: electronic conduction, oxygen vacancies ionic conduction, and mixed conduction of ions and holes. BLSF compounds' conduction mechanisms are still indistinct. Various mechanisms have been put forward to explain the conductivity property. These mechanisms may be very correct according to the corresponding experiments, but none of them is widely accepted.

Bi_3TiNbO_9 (BTN), which is made up of $(Bi_2O_2)^{2+}$ layers between which two $(BiTiNbO_7)^{2-}$ layers (m = 2) are inserted, has a very high T_C of ~914 °C. As a sensing material, BTN is promising for fabricating piezoelectric accelerometers with operating temperature above 500 °C [8], which can be used for the high-temperature vibration monitoring of some large power equipment such as aircraft engines, gas turbines, power generators, etc. However, the piezoactivity of pure BTN ceramics is very low ($d_{33} \leq 7$ pC/N) [9]. The resistivity of BTN is only about 10^7 Ω·cm at 400 °C, for example [10]. Up to now, most reported studies about BTN have concentrated on its crystal structure [11–14], the electrical properties of pure BTN ceramics [15–18], and improvement in its piezoelectric ability [19–21]. For example, S.V. Zubkov doped Gd elements in BTN, which can increase the Curie temperature to 950 °C [22]. Gd element was also found to provide high insulation and low loss [23]. Chen et al. codoped W/Cr into $Bi_4Ti_3O_{12}$ of BLSFs, which increased the resistivity (σ_{dc} (600 °C)) and piezoelectric properties (d_{33} (RT)) to 2.94×10^6 Ω·cm and 28 pC/N, respectively [24,25]. In addition, Chen et al. codoped Mo/Cr into $CaBi_2Nb_2O_9$, which increased the multifaceted performances of $CaBi_2Nb_2O_9$ (d_{33} = 15 pC/N, T_C = 939 °C, σ_{dc} (600 °C) = 3.33×10^5 Ω·cm) [26]. However, there has been no significant improvement of the piezoelectric properties of BTN-based ceramics. There is also no clear understanding of their conduction behavior.

In this work, we studied the effects of Gd/Cr codoped on the microstructure, AC conduction mechanisms, and electrical impedance spectrum of BGTN−0.2Cr ceramic, focusing on the effect of grain size on conduction mechanism and impedance spectroscopy. A type of piezoelectric ceramic that can be used as a sensing material for piezoelectric sensors with operating temperatures above 600 °C was developed in this work.

2. Experimental Section

2.1. Sample Preparation

A kind of Gd/Cr codoped Bi_3TiNbO_9 Aurivillius phase ceramics, with the formula of $Bi_{2.8}Gd_{0.2}TiNbO_9$ + 0.2 wt% Cr_2O_3 (abbreviated as the BGTN−0.2Cr ceramic hereafter), was prepared by using the conventional solid-state reaction route in two steps. First, the metal oxides Bi_2O_3 of 99% purity, Gd_2O_3 of 99.99% purity, TiO_2 of 99% purity, and Nb_2O_5 of 99.99% purity (as raw materials) were weighed according to the stoichiometric ratio (Bi_2O_3, Gd_2O_3, TiO_2 and Nb_2O_5 produced in Chron Chemicals, Chengdu, China; Cr_2O_3 produced in Aladdin, Shanghai, China). These raw materials were mixed evenly by ball milling for 6 h, using alcohol as solvent and zirconia balls as grinding media. The dried mixture was calcined at 850 °C for 4 h to synthesize the components of $Bi_{2.8}Gd_{0.2}TiNbO_9$. Second, the calcined powders, with x wt% Cr_2O_3 of 99.95% purity added, were ground for 12 h under the same grinding conditions and granulated with polyvinyl alcohol (abbreviated as PVA, produced in Chron Chemicals, Chengdu, China) as a binder. The powders were pressed into discs with a diameter of 10 mm and a thickness of 0.8 mm under an isostatic pressure of 10 MPa. After firing at 650 °C for 2 h to burn out the PVA, these discs were sintered in a sealed alumina crucible at 1100 °C for 2 h to obtain the dense ceramics. Finally, Ag electrodes were screen-printed on both surfaces of the polished ceramics and fired at 700 °C for 10 min. For comparison, the pure Bi_3TiNbO_9 ceramic was prepared by the same process.

2.2. Sample Characterization

The crystallographic structure of the samples was characterized by X-ray diffractometer (XRD, DX−2700B, HAOYUAN INSTRUMENT, Dandong, China) using CuK$_\alpha$ radiation. The microscopic morphology of the samples was observed by scanning electron microscope (SEM, Quanta FEG 250, FEI, Waltham, MA, USA). The elemental analysis of the samples was carried out by the energy-dispersive X-ray spectroscopy (EDS) attached to the scanning electron microscope. In the frequency range of 100 Hz–100 kHz, the dielectric constant and loss of the samples as a function of temperature (room temperature~650 °C) were detected by an LCR meter (TH2829A, Tonghui Electronic, Changzhou, China) attached to a programmable furnace. At various temperatures (room temperature~650 °C), the AC impedance spectrum of the samples as a function of frequency (20 Hz–2 MHz) was measured by using an impedance analyzer (TH2838, Tonghui Electronic) attached to a programmable furnace. The electromechanical resonance spectroscopy of the poled samples was investigated by a broadband LCR digital bridge in the frequency range from 280 to 390 kHz. The planar electromechanical coupling factors k_p, mechanical quality factors Q_m, and planar frequency constant N_p were determined according to Onoe's equations [27]. The piezoelectric charge coefficient of the samples was measured using a quasistatic d_{33}/d_{31} m (ZJ−6AN, IACAS, Beijing, China).

3. Results and Discussion

3.1. Phase Structure of Ceramics

Figure 1 shows the XRD analysis of the sintered ceramics. As can be seen from Figure 1a, both for the pure BTN ceramic and the BGTN−0.2Cr ceramic, their observed XRD patterns agree with the standard X-ray diffraction powder patterns from JCPDS card No. 39-0233 well. Therefore, both samples were identified as the pure Bi_3TiNbO_9 phase with orthorhombic structure and space group of $A2_1am$. No second phase was found in the XRD pattern of the BGTN−0.2Cr ceramic. The introduced Gd_2O_3 and Cr_2O_3 formed a complete solid solution with Bi_3TiNbO_9. The strongest diffraction peak of the BGTN−0.2Cr ceramic was the (1 × 1 × 5) peak, which is consistent with the rule of the strongest diffraction peak of BLSF ceramics (1 × 1 × 2m + 1) [28]. In view of the similar ionic radius, it is believed that Gd^{3+} (r = 0.0938 nm) entered the A-site (Bi^{3+}: r = 0.103 nm) of the perovskite unit while Cr^{3+} (r = 0.0615 nm) entered the B-site (Ti^{4+}: r = 0.0605 nm). Meanwhile, the stability of the ABO_3—type perovskite structure can be described by the tolerance factor (t), which can be expressed as [29]:

$$t = \frac{r_A + r_o}{\sqrt{2}(r_B + r_o)} \qquad (1)$$

where r_A, r_B, and r_o are the ionic radii of A, B, and the oxygen ion, respectively. The perovskite structure remains stable when t is between 0.77 and 1.10. However, the value of the tolerance factor may be further limited in BLSF due to the structural incompatibility between the pseudo perovskite layer and the bismuth-oxygen layer, which is caused by the mismatching of their transverse dimension. Subbarao [1] proposed that when m = 2, t is limited in the range of 0.81~0.93. We obtained t = 0.86 for BTN and t = 0.82 for BGTN−0.2Cr. The decrease in the tolerance factor could demonstrate the successful substitution of Gd^{3+} and Cr^{3+} for Bi^{3+} and Ti^{4+} at the A- and B-sites, respectively, in the perovskite unit of BTN.

The Rietveld refinement was performed for the XRD patterns of the pure BTN ceramic and the BGTN−0.2Cr; the refined factors and cell parameters are shown in Figure 1b,c. When compared with pure BTN, BGTN−0.2Cr showed a contracted unit cell as well as a smaller orthorhombic distortion (a/b). The substitution of Gd^{3+} for Bi^{3+} at the A—site tended to induce a principle change in the structural distortion of perovskite blocks composed of (Ti, Nb)O_6 octahedrons. However, the substitution of Cr^{3+} for Ti^{4+} at the B—site may have decreased the tilting angle of the (Ti, Nb)O_6 octahedron, leading to reduced distortion in the perovskite blocks.

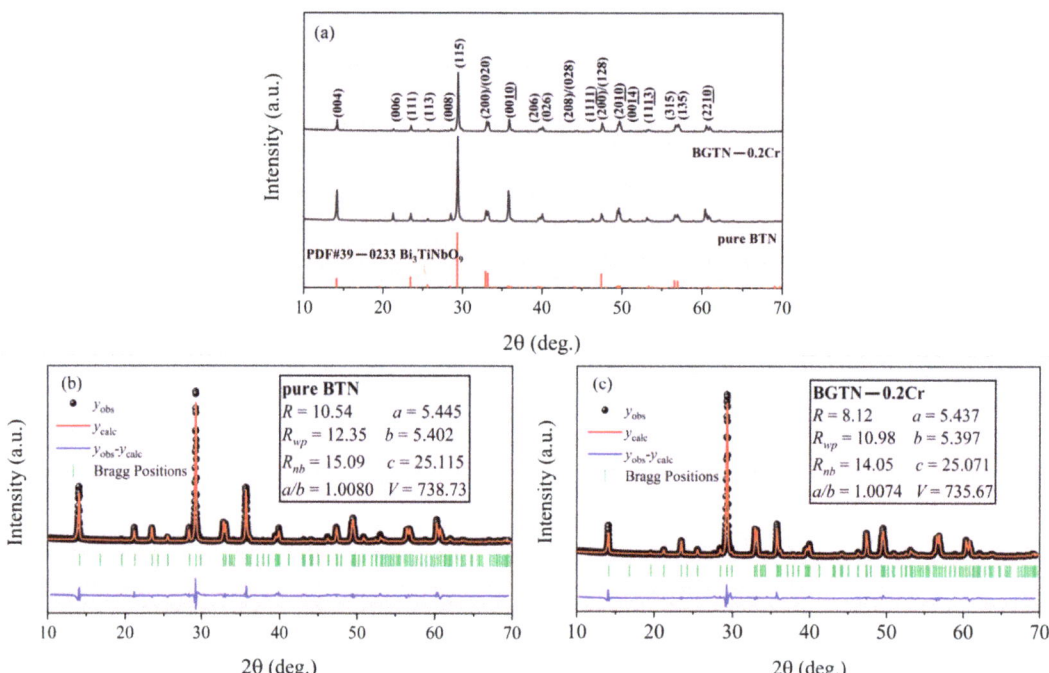

Figure 1. XRD analysis of the sintered ceramics: (**a**) observed XRD patterns of the pure BTN ceramic and the BGTN−0.2Cr ceramic; (**b**) Rietveld XRD refinement of the pure BTN ceramic; (**c**) Rietveld XRD refinement of the BGTN−0.2Cr ceramic.

3.2. Grain Morphology and Chemical Composition of Ceramics

Figure 2 shows the SEM and EDS analysis of the BGTN−0.2Cr ceramic, focused on its thermal-etched surface. As can be seen from the SEM image inserted in the EDS, a dense microstructure with well-defined grain boundaries was formed in the ceramic. All the grains were closely stacked, with random orientation, which agrees with the weaker intensity of the $(0 \times 0 \times l)$ diffracted peaks observed in Figure 1. The microstructure was composed of plate−like grains. Such grain growth possessed a high anisotropy such that the length (L) was larger than the thickness (T), which can be attributed to the higher grain growth rate in the direction perpendicular to the c-axis of the BLSF grains [30]. It is well known that the crystal grain aspect ratio (L/T) has a significant influence on the resistivity of BLSF ceramics. A higher aspect ratio is often related to higher resistivity [31]. The average thickness of these plate-like grains was 1.47 µm, while the length was about 8.93 µm. The high aspect ratio, with a L/T value of 6.1, was expected to lend higher resistivity to the BGTN−0.2Cr ceramic. Furthermore, EDS analysis showed that the BGTN−0.2Cr ceramic contained six elements: Bi, Gd, Ti, Nb, Cr, and O. All the elements presented a uniform distribution in the detected zone. Both gadolinium and chromium were successfully incorporated into the BTN.

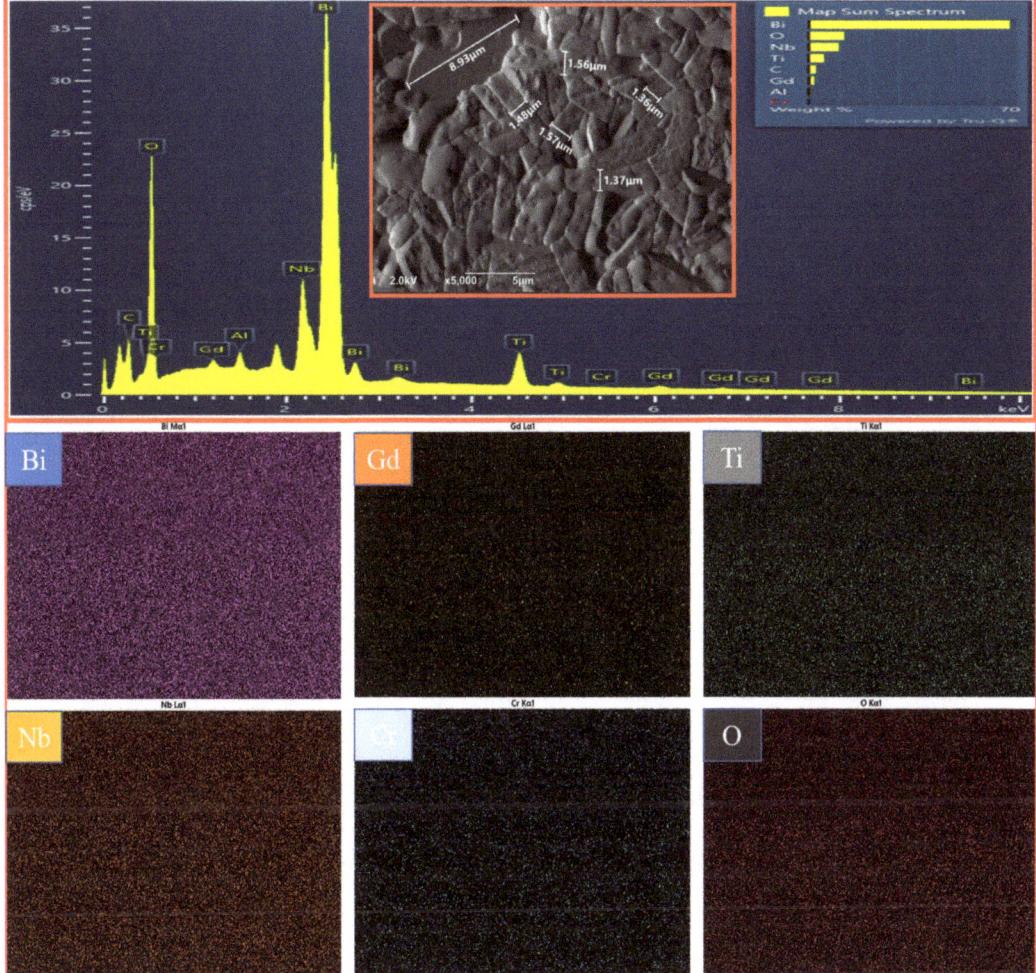

Figure 2. SEM and EDS analysis of the BGTN−0.2Cr ceramic, focused on its thermal-etched surface (the grain size was determined using the intercept procedure on the basis of the SEM image).

3.3. Electrical Conduction Behaviors of Ceramics

In order to further understand the conduction mechanism at high temperature, the results of electrical conduction spectroscopy of the BGTN−0.2Cr ceramic at high temperature are shown in Figure 3. It can be seen from Figure 3a that the conductivity of the BGTN−0.2Cr ceramic did not change with frequency in the low−frequency section at various temperatures, while in the high-frequency section, the conductivity increased with the increase in frequency. This is consistent with the jump relaxation model proposed by Funke [32]. In the low-frequency section, the migration of charge carriers was mainly implemented through long-distance jump, which led to direct current conductivity. As the frequency increased, the mobility of charge carriers was gradually limited, and the conductivity became positively related to frequency. The functional relationship between conductivity and frequency is consistent with the general Jonscher's theory [33]:

$$\sigma(\omega) = \sigma_{dc}(T) + A\omega^n \quad (n = 0 \sim 1) \quad (2)$$

where $\sigma(\omega)$ is the total conductivity, $\sigma_{dc}(T)$ is the dc conductivity, A is a constant with temperature dependence, ω is the angular frequency, n is the frequency index factor, and $A\omega^n$ represents the ac conductivity. $\sigma_{dc}(T)$, n, and A can be fitted by Equation (2). Based on Jonscher's theory, the frequency dependence of ac conductivity originated from the relaxation of the ionic atmosphere after the movement of charge carriers, as shown in Figure 3a.

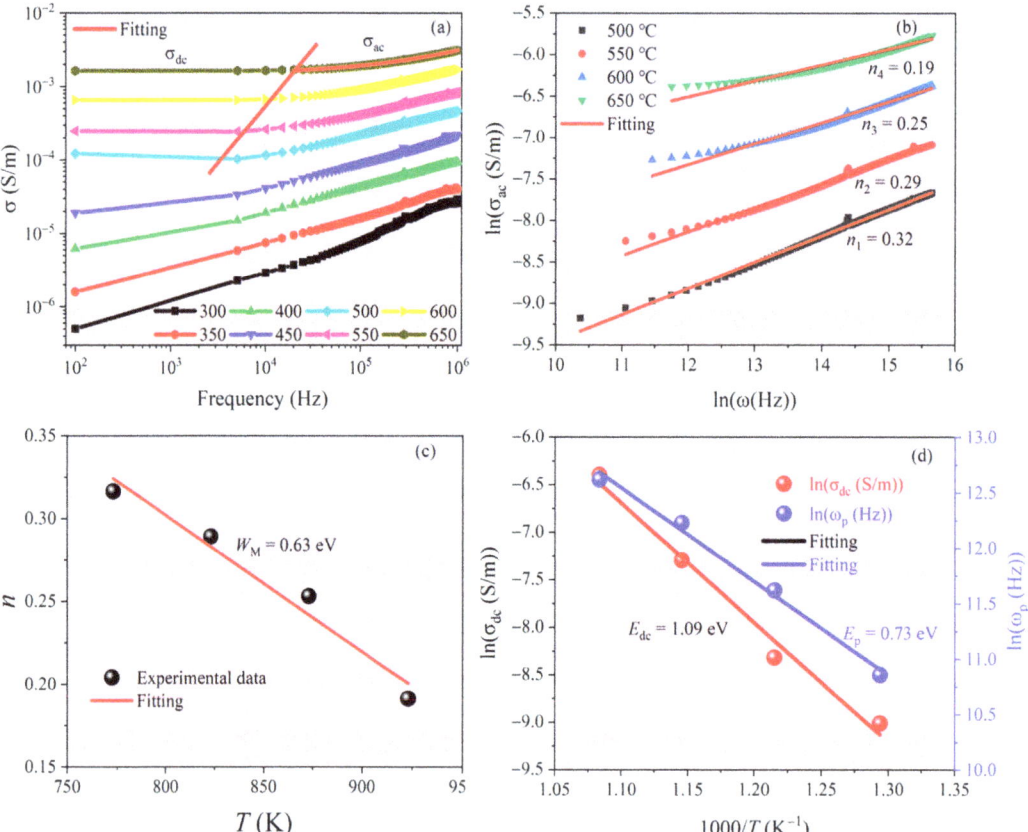

Figure 3. Electrical conduction spectroscopy of the BGTN−0.2Cr ceramic at high temperature: (**a**) frequency dependence of the total conductivity; (**b**) relationship between ac conductivity and angular frequency; (**c**) values of the frequency index factor n calculated by the general Jonscher's theory; (**d**) Arrhenius fitting of the plots of $\ln\sigma_{dc}$ and $\ln\omega_p$ vs. $1000/T$.

The changes in the temperature-related frequency index factor n provide information for the origins of conductance. The value of n in the correlated barrier–hopping (CBH) model decreases with temperature rising [34]. The results obtained for the BGTN−0.2Cr ceramic are shown in Figure 3b,c and are well in line with the CBH model. The first-order approximation of the frequency index factor n of the CBH model is given as Equation (3) [35]:

$$n = 1 - \frac{6K_B T}{W_M} \quad (3)$$

where W_M is the maximum barrier height. The calculated W_M (~0.63 eV) was slightly less than the oxygen vacancy's activation energy (~0.6–1.2 eV), indicating that oxygen vacancies were the main charge carriers between local states. This may be related to the segmental ionization of the first- and second-order oxygen vacancies, which can be expressed by Equations (4) and (5), respectively:

$$V_O \leftrightarrow V_{\dot{O}} + e' \qquad (4)$$

$$V_{\dot{O}} \leftrightarrow V_{\ddot{O}} + e' \qquad (5)$$

where $V_{\dot{O}}$ and $V_{\ddot{O}}$ are the first- and second-stage ionized oxygen vacancies, respectively. In many BLSF materials, the activation energies of the first- and second-stage ionized oxygen vacancies (E_I and E_{II}) have been reported as ~0.5 eV and ~1.2 eV, respectively [36]. For instance, E_I and E_{II} in SBTW0.04 are 0.57 eV and 0.74 eV [37], and E_{II} in $Bi_{2.8}Nd_{0.2}NbTiO_9$ is 0.9 eV [38]. The conductive and relaxation behaviors related to these BLSF materials can be attributed to long-range/local migration of two-stage ionized oxygen vacancies.

The conductivity of ion conductive materials is related not only to movable ion concentrations but also to ion jump frequencies [39]. The hopping angular frequency ω_p of ac conductivity can be fitted by Equation (6), and the activation energy of dc conduction and hopping conduction can be determined and calculated by Arrhenius fitting shown in Equations (7) and (8):

$$\omega_p = \left(\frac{\sigma_{ac}}{A}\right)^{1/n} \qquad (6)$$

$$\omega_p = \omega_0 \exp(-E_p/kT) \qquad (7)$$

$$\sigma_{dc} = \sigma_0 \exp(-E_{dc}/kT) \qquad (8)$$

where ω_0 and σ_0 are pre−exponential factors, k is the Boltzmann constant, E_p is the activation energy of hopping conduction, and E_{dc} is the dc conduction activation energy. The calculated E_{dc} (~1.09 eV) in Figure 3d was slightly smaller than the activation energy (~1.2 eV) of the second ionized oxygen vacancies, which indicates that $V_{\dot{O}}$ and $V_{\ddot{O}}$ were involved in conduction during the dc conductance process. Furthermore, the reduction of E_p (0.73 eV) indicates that when the oxygen vacancies migrated from long- to short-range jumps, the activation energy decreased, which may have increased the carrier mobility during the ac conduction process. E_p was slightly greater than W_M, indicating that carriers were over the barrier height and then had a short-distance jump participation in the ac conduction process. The higher E_p may be also caused by relaxation.

When poled ferroelectric ceramics are used as piezoelectric elements, over time, the polarization change caused by the applied stress is offset by charge movement caused by internal conduction inside the material. At high frequencies, the charge compensation caused by conductivity can be ignored, because the change rate of charge caused by the applied stress is much faster than the time constant (RC). However, at low frequencies, signals from sensors or generators may be significantly attenuated. The minimum useful frequency or lower limit frequency (f_{LL}) is inversely proportional to the time constant, which can be calculated by Equation (9):

$$f_{LL} = \frac{1}{2\pi RC} = \frac{\sigma}{2\pi \varepsilon'} \qquad (9)$$

where σ is the dc conductivity, which can be deduced from AC fields below 100 Hz, and ε' is the real part of the complex dielectric permittivity. It is well known that the RC time constant of BLSF ceramics tends to become very low at high temperature because of the sharp decline of resistivity [40]. In line with Equation (9), the values of f_{LL} in the temperature range of 450–650 °C were calculated for the BGTN−0.2Cr ceramic and are shown in Figure 4. The values of f_{LL} showed a sharp rise when the temperature increased from 450 to 500 °C, and then the increase in f_{LL} with temperature began to slow down when the temperature exceeded 500 °C. In addition to the significant decrease in resistance with temperature, the capacitance of the ferroelectric material is closely related to the temperature. Therefore, the temperature correlation of f_{LL} combines the effects of capacitance and resistance, which can be considered as a useful quality factor for evaluating the service performance of ferroelectric materials.

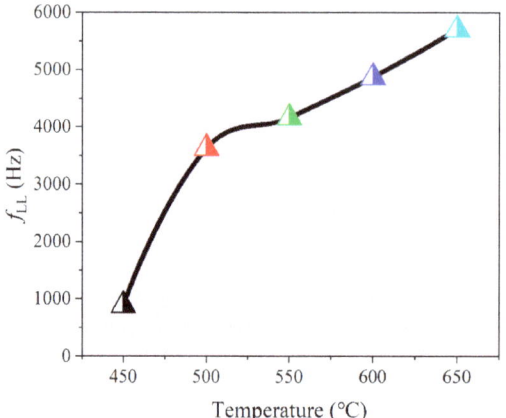

Figure 4. Lower limit frequency (f_{LL}) of the BGTN−0.2Cr ceramic as a function of temperature.

3.4. Electrical Impedance Spectroscopy of Ceramics

In order to study electrical behavior and distinguish the contribution of grains and grain boundaries to the conductivity of BGTN−0.2Cr ceramic, we analyzed the complex impedance data of BGTN−0.2Cr ceramic, considering both the real (Z') and imaginary (Z'') parts. The correlation function relationship can be expressed as follows:

$$Z = Z' - jZ'' \tag{10}$$

$$Z' = \frac{R\left(1 + (\omega\tau)^{1-\alpha}\cos\left[(1-\alpha)\frac{\pi}{2}\right]\right)}{1 + (\omega\tau)^{2(1-\alpha)} + 2(\omega\tau)^{1-\alpha}\cos\left[(1-\alpha)\frac{\pi}{2}\right]} \tag{11}$$

$$Z'' = \frac{R(\omega\tau)^{1-\alpha}\sin\left[(1-\alpha)\frac{\pi}{2}\right]}{1 + (\omega\tau)^{2(1-\alpha)} + 2(\omega\tau)^{1-\alpha}\cos\left[(1-\alpha)\frac{\pi}{2}\right]} \tag{12}$$

where $\tau = RC$ and $\alpha = 0\sim1$ is the proportion of the relaxation time distribution.

It can be seen from Figure 5a that the impedance value had a monotonous reduction as temperature and frequency rose in the low frequency range ($\leq$10 kHz). The reduction of Z' as the temperature rose in the low−frequency part suggests that conductivity increased with temperature. It was also seen that while Z' decreased as frequency increased, after reaching a fixed frequency ($\geq$200 kHz), the value of Z' became higher as the temperature increased and merged when the frequency increased further. This sudden change represents a possibility that the conductivity increased as frequency and temperature rose because of the fixed carriers at low temperatures and the defects at high temperatures. The combination of Z' at all temperatures at high frequencies may be due to the release of space charges, which would have led to a decrease in the resistance of the material. Figure 5b shows the change of Z'' with frequency at different temperatures. Obviously, as the frequency increased, Z'' reached a maximum value, which points to the relaxation process in the system. As the temperature increased, the maximum value shifted to higher frequencies. This shows that the relaxation was related to both temperature and frequency. The appropriate temperature activated the particles to cause large motions, and the appropriate frequency caused resonance. When the temperature matched the frequency, the relaxation phenomenon induced was the most obvious (Z'' maximum). It is well known that in perovskite−type compounds, the short-range motion of oxygen vacancies is a common phenomenon that contributes to high-temperature relaxation [41]. Because of the dispersion of bulk grains, Z'' merged at higher frequencies, which signified that the space charge was released. At the same time, the peak value decreased as the temperature increased and tended to become wider. The widening of the peak at higher temperatures indicates the existence of temperature-dependent relaxation. The substances causing relaxation of the material at high temperature may be vacancies or defects [42].

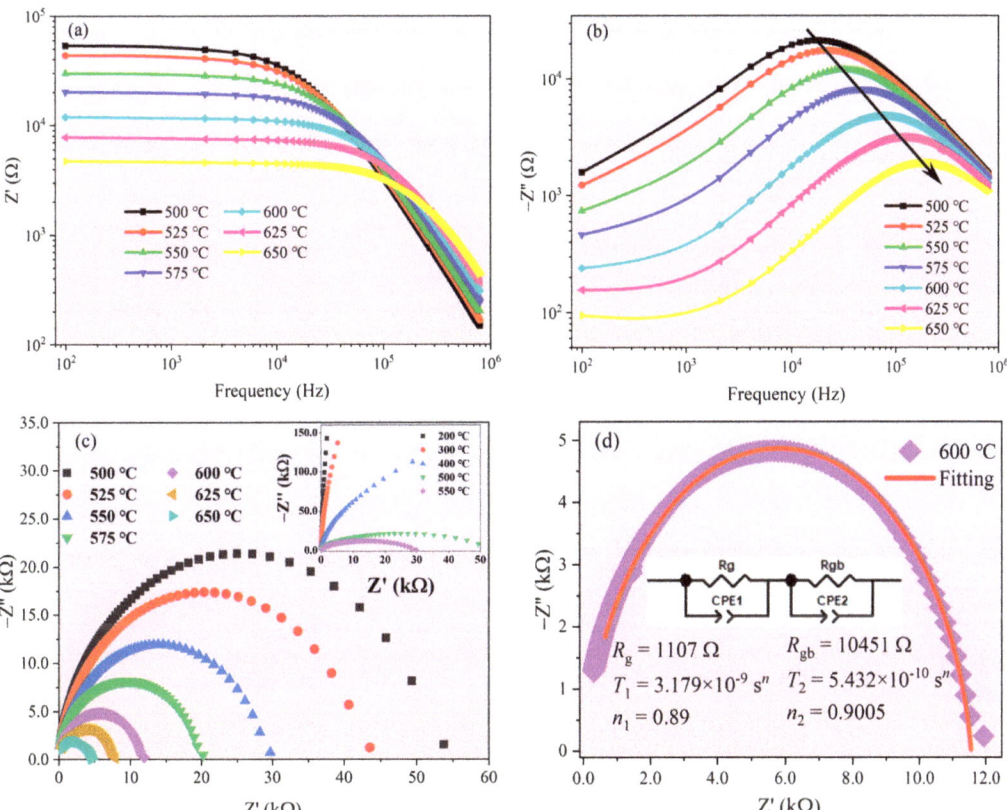

Figure 5. Electrical impedance spectroscopy of the BGTN−0.2Cr ceramic at high temperature: (**a**) Z'; (**b**) $-Z''$; (**c**) $-Z''$ vs. Z' (inset shows the impedance curve measured below 550 °C); (**d**) fitting for the Nyquist plot at 600 °C.

The Nyquist plot (Z'' vs. Z') from 200 to 650 °C is shown in Figure 5c. The impedance curve at low temperature (~200 °C) was close to the y-axis (imaginary part), which shows the high insulation properties of the ceramic at low temperatures. As the temperature increased, the impedance curve deviated from the y-axis and gradually curved toward the x-axis (real part) to form a deformed, asymmetric, semicircular arc. The asymmetry from the ideal semicircular arc suggests the possibility of multiple relaxation behaviors in BGTN−0.2Cr ceramic. The temperature continued to rise, and the radius of the deformed semicircular arc gradually decreased, indicating that the resistance of the ceramic gradually decreased as the temperature increased.

In order to judge whether multiple relaxation processes existed in BGTN−0.2Cr ceramic, the complex impedance spectrum was analyzed by z-view simulation software. The results are shown in Figure 5d. In the illustration, a resistor R_g and a constant phase element Q_g (CPE) connected in parallel represent the contribution of the crystal grain, and the other parallel element (R_{gb} and Q_{gb}) represents the contribution of the grain boundary. The total resistance of this equivalent circuit can be expressed as:

$$Z^* = \frac{R_g}{\left[1 + R_g T_1 (i\omega)^{n_1}\right]} + \frac{R_{gb}}{\left[1 + R_{gb} T_2 (i\omega)^{n_2}\right]} \tag{13}$$

where R_g, R_{gb}, T_1, and T_2 are the resistance and variables related to the relaxation time distribution from the grain and grain boundaries, respectively; n (0~1) is the distribution of relaxation time; and $n = 1$ is the ideal Debye relaxation response. The fitting curve simulated by the Z-View software matched our experimental data well, which confirmed that two kinds of relaxation mechanisms were involved in the BGTN−0.2Cr ceramic, i.e., grain boundaries contributed to the relaxation at low frequency, while grains contributed at high frequency [43]. The results of the simulation analysis showed that the grain boundary resistance (R_{gb}) was far greater than the bulk grain resistance (R_g) as shown in Figure 5d. This means that at high temperatures, the concentration of oxygen vacancies and captured electronics in the grain boundary was lower. n_2 was greater than n_1, which indicates that the impedance behavior at the grain boundaries was closer to the ideal Debye relaxation model.

The temperature dependence of R_g and R_{gb} of the BGTN−0.2Cr ceramic is displayed in Figure 6. The rate at which the grain's resistance decreased as the temperature rose was slower than that of the grain boundary. The calculation of resistivity and its temperature dependence can be described by the following formulas:

$$\rho = \frac{RS}{L} \tag{14}$$

$$\rho = \rho_0 \exp(E_{con}/kT) \tag{15}$$

where E_{con} is the activation energy for conduction, ρ_0 is the pre-exponential factor, and k is the Boltzmann constant. As shown in Figure 6, as the temperature rose, E_g increased from 0.49 to 1.15 eV, while E_{gb} increases from 0.64 to 1.34 eV. Such a significant increase in the activation energy confirmed that the charge carriers responsible for the electrical conduction process at high temperature changed from the primary−ionized oxygen vacancies to the secondary-ionized ones. No matter what kind of charge carriers dominated the electrical conduction, E_g was always lower than E_{gb}, which implies that the carrier concentration or migration speed of the grains was higher than that of the grain boundaries.

3.5. Electrical Modulus Spectroscopy of Ceramics

The impedance spectrum data emphasized only the maximum resistance element of microscopic components. In order to better understand the relaxation behaviors in BGTN−0.2Cr ceramic, electrical modulus analysis and impedance analysis complemented each other. The modulus spectrum handles the minimum capacitive element of the microscopic components and can suppress electrode interface effects [44,45]. Physically, the electrical modulus corresponds to the relaxation of the electric field in the material when the electric displacement remains constant. Therefore, the electrical modulus represents the real dielectric relaxation process, which can be expressed as:

$$M = M' + j M'' = j\omega C_0 Z \tag{16}$$

$$M' = \omega C_0 Z'' \tag{17}$$

$$M'' = j\omega C_0 Z\prime \tag{18}$$

where C_0 is the capacitance of free space given by $C_0 = \varepsilon_0 A/d$ [46].

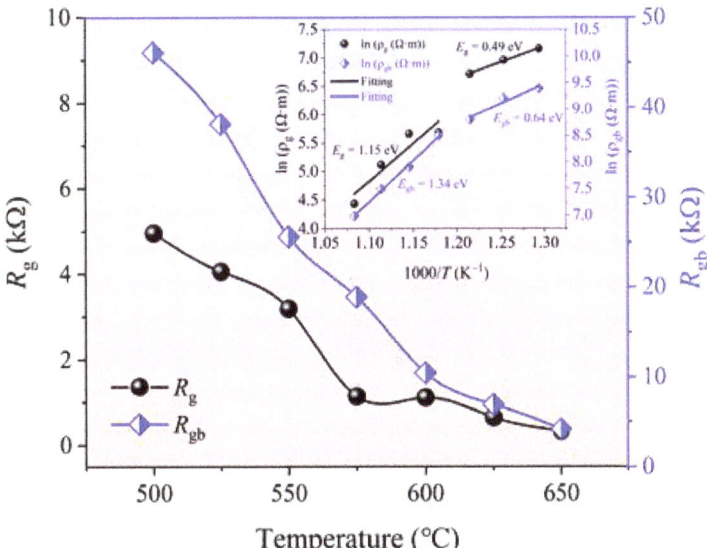

Figure 6. Temperature dependence of R_g and R_{gb} of the BGTN−0.2Cr ceramic (inset shows the Arrhenius fitting of the plots of resistivity vs. temperature).

In Figure 7, the electrical modulus spectroscopy of the BGTN−0.2Cr ceramic at high temperature is shown as a function of the frequency with temperature changing. It is obvious from Figure 7a that every temperature showed an identical trend—as frequency rose, M' values increased, gradually slowing (M' gradually in a fixed value at higher frequencies). M' showed asymmetry because of the tensile index characteristics of the relaxation time of materials. The monotonic dispersion in the low-frequency area may be due to the short-range jump of the carriers. This result may be related to the lack of recovery power of the charge carrier migration under the control of the polarization electric field [47]. On the other hand, M'' increased as frequency rose and reached a peak of relaxation because of the bulk grain and grain boundary behaviors. In physics, the electrical modulus peak can determine the area that the charge carrier can be migrated long distance. The asymmetrical and wide M'' peaks imply that the nonexponential behaviors of the grain and grain boundary relaxation deviated from Debye-type relaxation. The behavior indicates that ion migration occurred by jumping accompanying the corresponding time-dependent mobility of other nearby charge carriers [48]. These relaxation peaks moved towards higher frequencies as the temperature increased. With the equation $\omega\tau = 1$, we can obtain the relaxation time τ_0' related to the electrical modulus and perform Arrhenius fitting to τ_0'. Comparing the activation energies obtained from the impedance and modulus spectra (Figure 8), we found that the two processes had similar activation energies and pre-exponential factors. This indicates that the two had a common relaxation mechanism—both were dominated by similar carriers. The grain boundaries contributed to low-frequency relaxation, and the grains contributed to high-frequency relaxation.

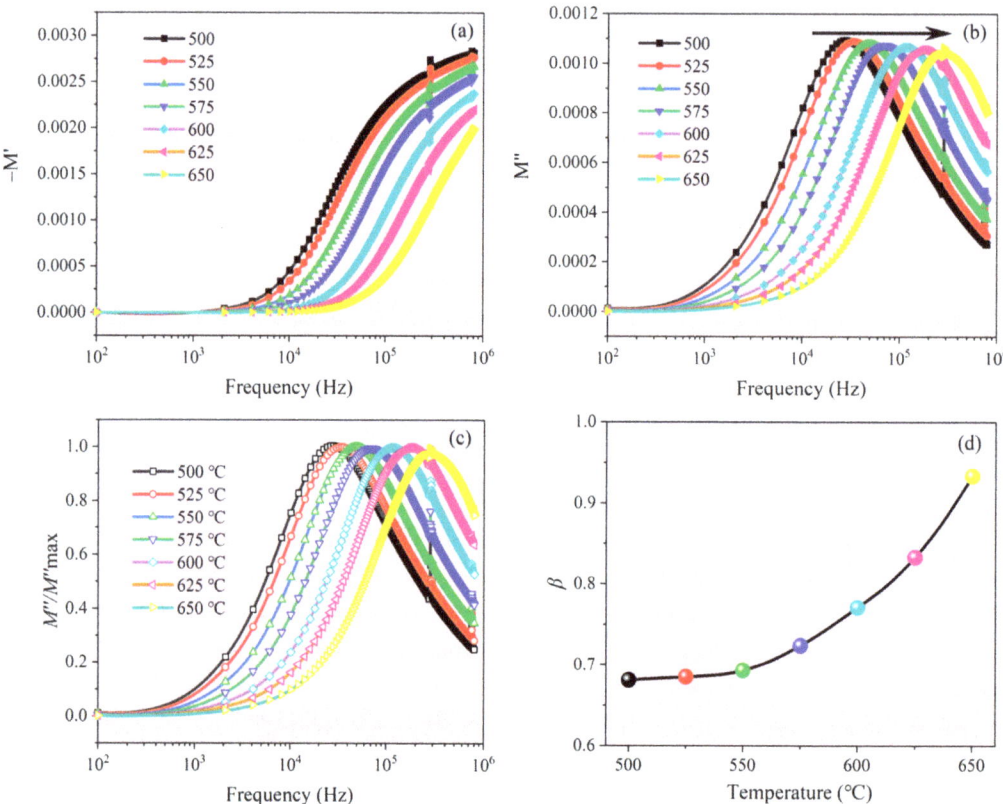

Figure 7. Electrical modulus spectroscopy of the BGTN−0.2Cr ceramic at high temperature: (**a**) −M'; (**b**) M''; (**c**) M''/M''_{max}; (**d**) temperature dependence of β.

The M'' curves at different temperatures were normalized to a master curve with the peak position (f_{max}) and the peak height (M''_{max}) to research the relaxation process (Figure 7c). The shape of this peak was asymmetrical and wider than Debye-type relaxation. This phenomenon is well described by the Bergman formula [49]:

$$M''(\omega) = \frac{M''_{max}}{1 - \beta + \frac{\beta}{1+\beta}[\beta(\omega_{max}/\omega) + (\omega/\omega_{max})]^\beta} \quad (19)$$

where M''_{max} is the maximum value of M'', ω_{max} is the angular frequency corresponding to M''_{max}, and β indicates the extent of deviation from the ideal Debye model—the smaller the value of β, the greater the deviation from Debye−type relaxation ($\beta = 1$, see [48]). β increased as temperature rose (Figure 7d), which proves that the relaxation behavior of the BGTN−0.2Cr ceramic grew closer to Debye−type relaxation at higher temperature.

In Figure 8, the frequency dependence of Z''/Z''_{max} and M''/M''_{max} at 575 °C is shown. It was discovered that the impedance and electrical modulus each showed only one peak of relaxation behaviors. Because of the different focuses of the impedance and modulus, the impedance spectrum with the more resistive grain-boundary component displayed only a single peak, while the single peak of M''/M''_{max} may be due to the contribution of the grains. The Z''/Z''_{max} peak appeared at a lower frequency, which suggests that the grain boundary gathered a large amount of oxygen vacancies and space charges. The M''/M''_{max} peak appeared at a higher frequency, which suggests that the grain has a carrier concentration or migration speed below the grain boundary. E_a dominated by Z was slightly larger than $E_a{'}$ dominated by M, which once again proves that the activation energy of the grain boundary was slightly larger than that of the grain, as shown in Figure 6.

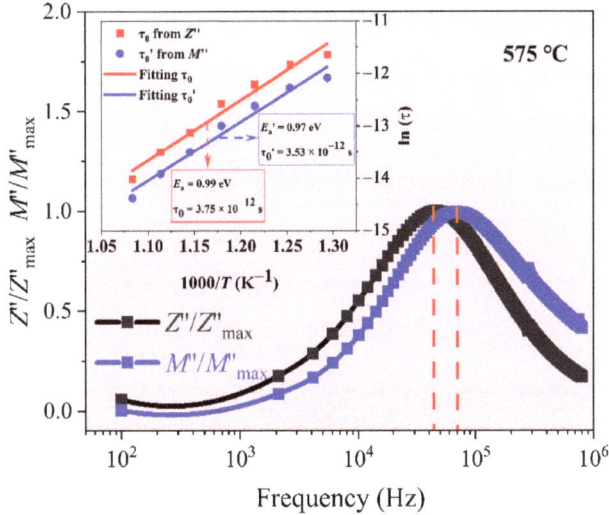

Figure 8. Comparison between Z''/Z''_{max} peak and M''/M''_{max} peak for the BGTN−0.2Cr ceramic (inset shows the Arrhenius fitting of the plots of relaxation time vs. temperature).

3.6. Electromechanical Resonance Spectroscopy of Ceramics

Figure 9 shows the electromechanical resonance spectroscopy results for the BTGN−0.2Cr ceramic at room temperature. This measured the frequency dependence of impedance |Z| and the phase angle θ of the piezoceramic as a resonator. The paired peak of resonance-antiresonance around 285.5 kHz showed that the sample was in the radial-extensional vibration mode. In an ideal poling state, piezoelectric materials should exhibit an impedance phase angle θ approaching 90° in the frequency range between the resonance (f_r) and antiresonance frequencies (f_a). As can be seen, the maximum phase angle θ_{max} was only 36.4°, indicating the insufficient poling of the sample. Since BTN has a high coercive electric field, as well as a low resistivity, a complete domain switching is difficultly achieved by the poling electric field. Things that give satisfaction is that the d_{33} value of the BTGN−0.2Cr ceramic has reached 18 pC/N, which provides it with a large competitiveness used as sensing materials for high−temperature piezoelectric sensors.

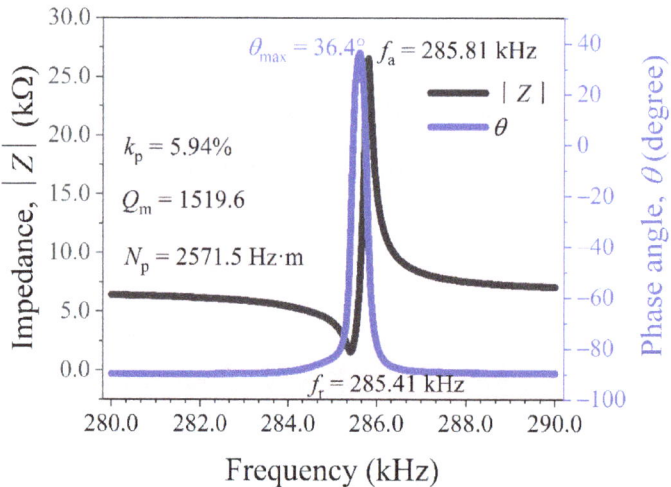

Figure 9. Electromechanical resonance spectroscopy of the BGTN−0.2Cr ceramic at room temperature.

4. Conclusions

An Aurivillius phase ceramic with a formula of $Bi_{2.8}Gd_{0.2}TiNbO_9 + 0.2$ wt% Cr_2O_3 (abbreviated as BGTN−0.2Cr) was prepared by the conventional solid−state reaction route. Both the microstructures and electrical conduction behaviors of the BGTN−0.2Cr ceramic were studied. The BGTN−0.2Cr ceramic was crystallized in a pure Bi_3TiNbO_9 phase and composed of plate-like grains. Its ac conduction behavior could be explained by the Funke's jumping relaxation model. The maximum barrier height W_M, hopping conduction activation energy E_p, and dc conduction activation energy E_c were determined to have values of 0.63 eV, 1.09 eV, and 0.73 eV, respectively. Impedance analysis in combination with modulus calculation revealed that grains provided a larger contribution to the conductance at high temperature and that β grew larger as temperature rose. The values of the activation energy (~1 eV) calculated from both Z'' and M'' peaks suggested the electrical relaxation process to be dominated by the thermal activation of oxygen vacancies as defect charge carriers. Moreover, the BGTN−0.2Cr ceramic had a high d_{33} of 18 pC/N as well as a high σ_{dc} of 1.52×10^5 Ω cm (600 °C). Such excellent electrical properties made it a competitive candidate for high−temperature piezoelectric materials.

Author Contributions: H.Z. conceived and designed the experiments; S.W. and D.W. performed the experiments; H.Z. analyzed the data; Q.C. contributed reagents/materials/analysis tools; H.Z. wrote the paper; Y.C. revised the paper. All authors have read and agreed to the published version of the manuscript.

Funding: This work was supported by Applied Basic Research of Sichuan Province (No. 2020YJ0317), the Open project from the Fujian Key Laboratory of Special Advanced Materials (Development of Bi_3TiNbO_9 High-temperature Piezoceramics, 2021), and the Open project from the Key Laboratory of Deep Earth Science and Engineering, Ministry of Education (DESE202007).

Institutional Review Board Statement: Not applicable.

Informed Consent Statement: Not applicable.

Data Availability Statement: The data presented in this study are available on request from the corresponding author.

Acknowledgments: The authors give thanks for SEM characterization technical support provided by the Institute of Advanced Study at Chengdu University.

Conflicts of Interest: The authors declare no conflict of interest. The authors alone were responsible for the content and writing of this article.

References

1. Subbarao, E.C. Crystal Chemistry of Mixed Bismuth Oxides with Layer-Type Structure. *J. Am. Ceram. Soc.* **1962**, *45*, 166–169. [CrossRef]
2. Aurivillius, B. Mixed Bismuth Oxides with Layer Lattices: I. *Arkiv Kemi* **1949**, *1*, 463–480.
3. Newnham, R.E.; Wolfe, R.W.; Dorrian, J.F. Structural basis of ferroelectricity in the bismuth titanate family. *Pergamon* **1971**, *6*, 1029–1039. [CrossRef]
4. Pardo, L.; Castro, A.; Millan, P.; Alemany, C.; Jimenez, R.; Jimenez, B. $(Bi_3TiNbO_9)_x(SrBi_2Nb_2O_9)_{1-x}$ aurivillius type structure piezoelectric ceramics obtained from mechanochemically activated oxides. *Acta Mater.* **2000**, *48*, 2421–2428. [CrossRef]
5. Hong, S.H.; Trolier-McKinstry, S.; Messing, G.L. Dielectric and Electromechanical Properties of Textured Niobium—Doped Bismuth Titanate Ceramics. *J. Am. Ceram. Soc.* **2000**, *83*, 113–118. [CrossRef]
6. De Araujo, C.P.; Cuchiaro, J.D.; McMillan, L.D.; Scott, M.C.; Scott, J.F. Fatigue—free ferroelectric capacitors with platinum electrodes. *Nature* **1995**, *374*, 627–629. [CrossRef]
7. Shulman, H.S.; Testorf, M.; Damjanovic, D.; Setter, N. Microstructure, Electrical Conductivity, and Piezoelectric Properties of Bismuth Titanate. *J. Am. Ceram. Soc.* **1996**, *79*, 3124–3128. [CrossRef]
8. Thompson, J.G.; Rae, A.D.; Withers, R.L.; Craig, D.C. Revised structure of Bi_3TiNbO_9. *Acta Crystallogr. Sect. B* **1991**, *47*, 174–180. [CrossRef]
9. Zhang, Z.; Yan, H.X.; Dong, X.L.; Wang, Y.L. Preparation and electrical properties of bismuth layer–structured ceramic Bi3NbTiO9 solid solution. *Mater. Res. Bull.* **2003**, *38*, 241–248. [CrossRef]
10. Moure, A.; Pardo, L.; Alemany, C.; Millán, P.; Castro, A. Piezoelectric ceramics based on Bi_3TiNbO_9 from mechanochemically activated precursors. *J. Eur. Ceram. Soc.* **2001**, *21*, 1399–1402. [CrossRef]
11. Wolfe, R.W.; Newnham, R.E.; Smith, D. Crystal structure of Bi_3TiNbO_9. *Ferroelectrics* **1972**, *3*, 1–7. [CrossRef]
12. Withers, R.L.; Thompson, J.G.; Rae, A.D. The crystal chemistry underlying ferroelectricity in $Bi_4Ti_3O_{12}$, Bi_3TiNbO_9, and Bi_2WO_6. *J. Solid. State. Chem.* **1991**, *94*, 404–417. [CrossRef]
13. Yuan, J.; Nie, R.; Chen, Q.; Xiao, D.; Zhu, J. Structural distortion, piezoelectric properties, and electric resistivity of A-site substituted Bi_3TiNbO_9-based high-temperature piezoceramics. *Mater. Res. Bull.* **2019**, *115*, 70–79. [CrossRef]
14. Ricote, J.; Pardo, L.; Moure, A.; Castro, A.; Millán, P.; Chateigner, D. Microcharacterisation of grain-oriented ceramics based on Bi_3TiNbO_9 obtained from mechanochemically activated precursors. *J. Eur. Ceram. Soc.* **2001**, *21*, 1403–1407. [CrossRef]
15. Zhang, Z.; Yan, H.X.; Dong, X.L.; Wang, Y.L. Structural and electrical properties of Bi_3NbTiO_9 solid solution. *J. Inorg. Mater.* **2003**, *18*, 1377–1380.
16. Lisińska-Czekaj, A.; Czekaj, D.; Gomes, M.J.M.; Kuprianov, M.F. Investigations on the synthesis of Bi_3NbTiO_9 ceramics. *J. Eur. Ceram. Soc.* **1999**, *19*, 969–972. [CrossRef]
17. Zhang, Z.; Yan, H.X.; Dong, X.L.; Xiang, P. Grain orientation effects on the properties of a bismuth layer-structured ferroelectric (BLSF) Bi_3NbTiO_9 solid solution. *J. Am. Ceram. Soc.* **2004**, *87*, 602–605. [CrossRef]
18. Moure, A.; Pardo, L. Microstructure and texture dependence of the dielectric anomalies and dc conductivity of Bi_3TiNbO_9 ferroelectric ceramics. *J. Appl. Phys.* **2005**, *97*, 084103. [CrossRef]
19. Peng, Z.H.; Yan, D.X.; Chen, Q.; Xin, D. Crystal structure, dielectric and piezoelectric properties of Ta/W codoped Bi_3TiNbO_9 Aurivillius phase ceramics. *Curr. Appl. Phys.* **2014**, *14*, 1861–1866. [CrossRef]
20. Bekhtin, M.A.; Bush, A.A.; Kamentsev, K.E.; Segalla, A.G. Preparation and Dielectric and Piezoelectric Properties of Bi_3TiNbO_9, $Bi_2CaNb_2O_9$, and $Bi_{2.5}Na_{0.5}Nb_2O_9$ Ceramics Doped with Various Elements. *Neorg. Mater.* **2016**, *52*, 557–563. [CrossRef]
21. Zhang, F.Q.; Wahyudi, O.; Liu, Z.F.; Gu, H. Preparation and electrical properties of a new–type intergrowth bismuth layer–structured $(Bi_3TiNbO_9)_1(Bi_4Ti_3O_{12})_2$ ceramics. *J. Alloys Compd.* **2018**, *753*, 54–59. [CrossRef]
22. Zubkov, S.V. Crystal structure and dielectric properties of layered perovskite—like solid solutions $Bi_{3-x}Gd_xTiNbO_9$ (x = 0, 0.1, 0.2, 0.3) with high Curie temperature. *J. Adv. Dielectr.* **2020**, *10*, 2060002. [CrossRef]
23. Groń, T.; Maciejkowicz, M.; Tomaszewicz, E.; Guzik, M.; Oboz, M.; Sawicki, B.; Pawlus, S.; Nowok, A.; Kukuła, Z. Combustion synthesis, structural, magnetic and dielectric properties of Gd3+—doped lead molybdato-tungstates. *J.Adv. Ceram.* **2020**, *9*, 255–268. [CrossRef]
24. Chen, Y.; Peng, Z.H.; Wang, Q.Y.; Zhu, J. Crystalline structure, ferroelectric properties, and electrical conduction characteristics of W/Cr co–doped $Bi_4Ti_3O_{12}$ ceramics. *J. Alloys Compd.* **2014**, *612*, 120–125. [CrossRef]
25. Chen, Y.; Liang, D.Y.; Wang, Q.Y.; Zhu, J. Microstructures, dielectric, and piezoelectric properties of W/Cr co-doped $Bi_4Ti_3O_{12}$ ceramics. *J. Appl. Phys.* **2014**, *116*, 074108. [CrossRef]
26. Chen, Z.N.; Zhang, Y.H.; Huang, P.M.; Li, X. Enhanced piezoelectric properties and thermal stability in Mo/Cr co-doped $CaBi_2Nb_2O_9$ high-temperature piezoelectric ceramics. *J. Phys. Chem. Solids* **2020**, *136*, 109195. [CrossRef]
27. Onoe, M.; Jumonji, H. Useful formulas for piezoelectric ceramic resonators and their application to measurement of parameters. *J. Acoust. Soc. Am.* **1967**, *41*, 974–980. [CrossRef]
28. Du, X.F.; Chen, I.W. Ferroelectric thin films of bismuth-containing layered perovskites: Part II, $PbBi_2Nb_2O_9$. *J. Am. Ceram. Soc.* **1998**, *81*, 3260–3264. [CrossRef]

29. Li, Z.; Yang, M.J.; Park, J.S.; Wei, S.-H.; Berry, J.J.; Zhu, K. Stabilizing Perovskite Structures by Tuning Tolerance Factor: Formation of Formamidinium and Cesium Lead Iodide Solid−State Alloys. *Chem. Mater.* **2016**, *28*, 284–292. [CrossRef]
30. Wang, C.M.; Wang, J.F.; Zhang, S.J.; Shrout, T.R. Electromechanical properties of A−site (LiCe)−modified sodium bismuth titanate ($Na_{0.5}Bi_{4.5}Ti4O_{15}$) piezoelectric ceramics at elevated temperature. *J. Appl. Phys.* **2009**, *105*, 094110. [CrossRef]
31. Peng, Z.H.; Chen, Q.; Chen, Y.; Xiao, D. Microstructure and electrical properties in W/Nb co−doped Aurivillius phase $Bi_4Ti_3O_{12}$ piezoelectric ceramics. *Mater. Res. Bull.* **2014**, *59*, 125–130. [CrossRef]
32. Jonscher, A.K. Dielectric relaxation in solids. *J. Phys. D Appl. Phys.* **1999**, *32*, R57. [CrossRef]
33. Jonscher, A.K. The 'universal' dielectric response. *Nature* **1977**, *267*, 673–679. [CrossRef]
34. Elliott, S.R. Temperature dependence of ac conductivity of chalcogenide glasses. *Philos. Mag. B* **1978**, *37*, 553–560. [CrossRef]
35. Mohamed, C.B.; Karoui, K.; Saidi, S.; Guidara, K.; Rhaiem, A.B. Electrical properties, phase transitions and conduction mechanisms of the $[(C_2H_5)NH_3]_2CdCl_4$ compound. *Phys. B Condens. Matter.* **2014**, *451*, 87–95. [CrossRef]
36. Singh, G.; Tiwari, V.S.; Gupta, P.K. Role of oxygen vacancies on relaxation and conduction behavior of $KNbO_3$ ceramic. *J. Appl. Phys.* **2010**, *107*, 064103. [CrossRef]
37. Nayak, P.; Badapanda, T.; Singh, A.K.; Panigrahi, S. Possible relaxation and conduction mechanism in W^{6+} doped $SrBi_4Ti_4O_{15}$ ceramic. *Ceram. Int.* **2017**, *43*, 4527–4535. [CrossRef]
38. Zhang, H.T.; Yan, H.X.; Reece, M.J. The effect of Nd substitution on the electrical properties of Bi_3NbTiO_9 Aurivillius phase ceramics. *J. Appl. Phys.* **2009**, *106*, 044106. [CrossRef]
39. Almond, D.P.; Duncan, G.K.; West, A.R. The determination of hopping rates and carrier concentrations in ionic conductors by a new analysis of ac conductivity. *Solid State Ion.* **1983**, *8*, 159–164. [CrossRef]
40. Turner, R.C.; Fuierer, P.A.; Newnham, R.E.; Shrout, T.R. Materials for high temperature acoustic and vibration sensors: A review. *Appl. Acoust.* **1994**, *41*, 299–324. [CrossRef]
41. Li, W.; Hao, J.; Du, J.; Fu, P.; Sun, W.; Chen, C.; Xu, Z.; Chu, R. Electrical properties and luminescence properties of $0.96(K_{0.48}Na_{0.52})(Nb_{0.95}Sb_{0.05})–0.04Bi_{0.5}(Na_{0.82}K_{0.18})_{0.5}ZrO_3$–xSm lead−free ceramics. *J. Adv. Ceram.* **2020**, *9*, 72–82. [CrossRef]
42. Sumi, S.; Raon, P.P.; Koshy, P. Impedance spectroscopic investigation on electrical conduction and relaxation in manganese substituted pyrochlore type semicon−ducting oxides. *Ceram. Int.* **2015**, *41*, 5992–5998. [CrossRef]
43. Rehman, F.; Li, J.B.; Dou, Y.K.; Zhang, J.S.; Zhao, Y.J.; Rizwan, M.; Khalid, S.; Jin, H.B. Dielectric relaxations and electrical properties of Aurivillius $Bi_{3.5}La_{0.5}Ti_2Fe_{0.5}Nb_{0.5}O_{12}$ Ceramics. *J. Alloys Compd.* **2016**, *654*, 315–320. [CrossRef]
44. Wong, Y.J.; Hassan, J.; Hashim, M. Dielectric properties, impedance analysis and modulus behavior of $CaTiO_3$ ceramics prepared by solid state reaction. *J. Alloys Compd.* **2013**, *571*, 138–144. [CrossRef]
45. Morrison, F.D.; Sinclair, D.C.; West, A.R. Characterization of Lanthanum-doped barium titanate ceramics using impedance spectroscopy. *J. Am. Ceram. Soc.* **2001**, *84*, 531–538. [CrossRef]
46. Mahamoud, H.; Louati, B.; Hlel, F.; Guidara, K. Impedance and modulus analysis of the $(Na_{0.6}Ag_{0.4})_2PbP_2O_7$ compound. *J. Alloys Compd.* **2011**, *509*, 6083–6089. [CrossRef]
47. Mandal, S.K.; Dey, P.; Nath, T.K. Structural, electrical and dielectric properties of $La_{0.7}Sr_{0.3}MnO_3$–$ErMnO_3$ multiferroic composites. *Mater. Sci. Eng. B* **2014**, *181*, 70–76. [CrossRef]
48. Mohanty, V.; Cheruku, R.; Vijayan, L.; Govindaraj, G. Ce-substituted lithium ferrite: Preparation and electrical relaxation studies. *J. Mater. Sci. Technol.* **2014**, *30*, 335–341. [CrossRef]
49. Zurbuchen, M.A.; Sherman, V.O.; Tagantsev, A.K.; Schubert, J.; Hawley, M.E.; Fong, D.D.; Streiffer, S.K.; Jia, Y.; Tian, W.; Schlom, G. Synthesis, structure, and electrical behavior of $Sr_4Bi_4Ti_7O_{24}$. *J. Appl. Phys.* **2010**, *107*, 024106. [CrossRef]

Article

Effects of Oxide Additives on the Phase Structures and Electrical Properties of SrBi$_4$Ti$_4$O$_{15}$ High-Temperature Piezoelectric Ceramics

Shaozhao Wang [1], Huajiang Zhou [1], Daowen Wu [1], Lang Li [2] and Yu Chen [1,*]

[1] School of Mechanical Engineering, Chengdu University, Chengdu 610106, China; wangsz0908@163.com (S.W.); zhouhj1996@163.com (H.Z.); wdw132812976202021@163.com (D.W.)
[2] Key Laboratory of Deep Earth Science and Engineering, Ministry of Education, Sichuan University, Chengdu 610065, China; lilang@scu.edu.cn
* Correspondence: chenyuer20023@163.com; Tel./Fax: +86-28-84616133

Citation: Wang, S.; Zhou, H.; Wu, D.; Li, L.; Chen, Y. Effects of Oxide Additives on the Phase Structures and Electrical Properties of SrBi$_4$Ti$_4$O$_{15}$ High-Temperature Piezoelectric Ceramics. *Materials* **2021**, *14*, 6227. https://doi.org/10.3390/ma14206227

Academic Editor: Andres Sotelo

Received: 31 August 2021
Accepted: 8 October 2021
Published: 19 October 2021

Publisher's Note: MDPI stays neutral with regard to jurisdictional claims in published maps and institutional affiliations.

Copyright: © 2021 by the authors. Licensee MDPI, Basel, Switzerland. This article is an open access article distributed under the terms and conditions of the Creative Commons Attribution (CC BY) license (https://creativecommons.org/licenses/by/4.0/).

Abstract: In this work, SrBi$_4$Ti$_4$O$_{15}$ (SBT) high-temperature piezoelectric ceramics with the addition of different oxides (Gd$_2$O$_3$, CeO$_2$, MnO$_2$ and Cr$_2$O$_3$) were fabricated by a conventional solid-state reaction route. The effects of oxide additives on the phase structures and electrical properties of the SBT ceramics were investigated. Firstly, X-ray diffraction analysis revealed that all these oxides-modified SBT ceramics prepared presented a single SrBi$_4$Ti$_4$O$_{15}$ phase with orthorhombic symmetry and space group of $Bb2_1m$, the change in cell parameters indicated that these oxide additives had diffused into the crystalline lattice of SBT and formed solid solutions with it. The SBT ceramics with the addition of MnO$_2$ achieved a high relative density of up to 97%. The temperature dependence of dielectric constant showed that the addition of Gd$_2$O$_3$ could increase the T_C of SBT. At a low frequency of 100 Hz, those dielectric loss peaks appearing around 500 °C were attributed to the space-charge relaxation as an extrinsic dielectric response. The synergetic doping of CeO$_2$ and Cr$_2$O$_3$ could reduce the space-charge-induced dielectric relaxation of SBT. The piezoelectricity measurement and electro-mechanical resonance analysis found that Cr$_2$O$_3$ can significantly enhance both d_{33} and k_p of SBT, and produce a higher phase-angle maximum at resonance. Such an enhanced piezoelectricity was attributed to the further increased orthorhombic distortion after Ti^{4+} at B-site was substituted by Cr^{3+}. Among these compositions, Sr$_{0.92}$Gd$_{0.053}$Bi$_4$Ti$_4$O$_{15}$ + 0.2 wt% Cr$_2$O$_3$ (SGBT-Cr) presented the best electrical properties including T_C = 555 °C, $\tan \delta$ = 0.4%, k_p = 6.35% and d_{33} = 28 pC/N, as well as a good thermally-stable piezoelectricity that the value of d_{33} was decreased by only 3.6% after being annealed at 500 °C for 4 h. Such advantages provided this material with potential applications in the high-stability piezoelectric sensors operated below 500 °C.

Keywords: SrBi$_4$Ti$_4$O$_{15}$; high-temperature piezoceramics; oxide additives; curie temperature; piezoelectric coefficient; ion substitution

1. Introduction

Piezoelectric ceramics, which are a kind of synthetic polycrystalline ferroelectric material, have been used as sensing materials for many electrical devices such as ultrasonic transducers, vibration sensors and multi-layer actuators [1–5]. In recent years, the concern for the environmental pollution and people's health highlights the main differences between the commercial lead-based piezoelectric ceramics (lead zirconate titanate-PZT) and the developed lead-free piezoelectric ceramics (such as calcium barium titanate-BTO, potassium sodium niobate-KNN, etc.) [6–8]. Bismuth-layered structure ferroelectrics (BLSFs, also called Aurivillius phase), with large spontaneous polarization and fatigue-free properties, are promising candidates for ferroelectric random access memories (FRAMs) [9]. The chemical formula of BLSFs can be described as $(Bi_2O_2)^{2+} (A_{m-1}B_mO_{3m+1})^{2-}$, where m delegates to the number of octahedral layers in the perovskite layer between the bismuth oxide layers and values normally from 1 to 5. Such as Bi_2WO_6 ($m = 1$, $T_C = 950\ °C$) [10], Bi_3TiNbO_9 ($m = 2$, $T_C = 904\ °C$) [11], $Bi_4Ti_3O_{12}$ ($m = 3$, $T_C = 675\ °C$) [12], $SrBi_4Ti_4O_{15}$ ($m = 4$, $T_C = 520\ °C$) [13], $Sr_2Bi_4Ti_5O_{18}$ ($m = 5$, $T_C = 267\ °C$) [14]. Because of high Curie temperature, BLSFs have received more and more attention due to the urgent need of high-temperature sensor, actuator, and transducer applications in recent years [15].

Among the Aurivillius family, $SrBi_4Ti_4O_{15}$ (SBT) captures an orthorhombic symmetry with space group *A21am* at room temperature, including four perovskite-like TiO_6 octahedron units stacked in between $(Bi_2O_2)^{2+}$ layers [16]. However, some disadvantages associated with SBT such as difficulty in polarization, high leakage current, volatilization of the bismuth during sintering, and a low density and undesirable properties caused by the random arrangement of plate-like crystal grains [17], such as a tenuous piezoelectric activity ($d_{33} \sim 10$ pC/N) [18]. Over the last few decades, a large number of investigations focused on the modification of the electrical properties of SBT piezoceramics through the ionic substitution at A-site [19–21] or B-site [22–24]. Cao et al. [25] found a large enhancement of piezoelectric properties ($d_{33} = 30$ pC/N) in Mn-modified (B-site) $SrBi_4Ti_4O_{15}$ as well as good thermal stability at elevated temperatures, while its T_C remains almost unchanged at ~530 °C. However, most reports on enhancing the ferroelectric and piezoelectric properties of SBT ceramics concentrate on A-site rather than B-site [26]. For example, the Curie temperature increased by doping SBT with Na^+ and Pr^{3+} at A-site [27]. The dielectric constant and loss decreased, whereas the Curie temperature increased when Na^+ and Nd^{3+} were substituted to A-site in SBT [17]. A-site cerium-modified $SrBi_4Ti_4O_{15}$ ceramics showed a high stability of dielectric properties [28]. On the other hand, some oxide compounds like Cr_2O_3, MnO_2 and U_3O_8 as additives have been also proved to play a notable effect on the physical and electrical properties of BLSF ceramics [29].

Among these modified SBT ceramics, the compositions with both high Curie temperature and high piezoelectric property at the same time were rarely reported, and most of the as-reported works focused on the modification of the SBT ceramics with one kind of element/oxide. In this work, two kinds of oxides (Gd_2O_3, CeO_2, MnO_2 and Cr_2O_3) were co-doped into the SBT ceramics, and after these the oxide-modified SBT piezoceramics were fabricated via a traditional solid-state reaction process; the effect of these oxides on the phase structure and electrical properties of SBT ceramics have been investigated in detail.

2. Materials and Methods

2.1. Sample Preparation

Highly purified metal oxides of $SrCO_3$ (99%), Bi_2O_3 (99%), TiO_2 (99%), CeO_2 (99.99%) and Gd_2O_3 (99.99%) were weighed according to the stoichiometric formula of three designed compositions: $SrBi_4Ti_4O_{15}$ (SBT), $Sr_{0.92}Gd_{0.053}Bi_4Ti_4O_{15}$ (SGBT) and $Sr_{0.96}Ce_{0.04}Bi_4Ti_4O_{15}$ (SCBT), respectively. These powders as starting materials were milled for 10 h under the condition of ethanol as a dissolvent and zirconium ball as a milling medium. The dried mixtures were calcined at 850 °C for 4 h, and then the calcined powders of SGBT and SCBT were divided into four equal parts according to their quality, respectively. Secondly, 0.2 wt% of CeO_2 (99.99%), MnO_2 (99%) and Cr_2O_3 (99%) were severally added into three parts of SGBT powders, while 0.2 wt% of Gd_2O_3 (99.99%), MnO_2 (99%) and Cr_2O_3 (99%) were severally added into three parts of SCBT powders. Then, the mixtures of SBT, SGBT, SGBT-Ce, SGBT-Mn, SGBT-Cr, SCBT, SCBT-Gd, SCBT-Mn, SCBT-Cr were milled again in the same requirement. After drying, polyvinyl alcohol (PVA) as binder was added to the uniform mixture to form granules. The granules were pressed into pellets of 10 mm in diameter and 1 mm in thickness. After burning out PVA at 600 °C for 2 h, these pellets were sintered between 1100 °C and 1200 °C for 2 h in a sealed alumina crucible to obtain the ceramics with the maximum density.

2.2. Sample Characterization

The crystallographic structure of all sintered samples was determined by an X-ray diffractometer (DX2700, Dandong, China) employing Cu-Kα radiation (λ = 1.5418 Å). Meanwhile, the relative density of all sintered samples was calculated as the ratio of the apparent density measured by the Archimedes method to the theoretical density obtained from crystallographic structures (XRD). In order to measure the electrical properties, the samples were polished and fired with silver paste as the electrodes at 700 °C for 10 min. The dielectric constant (ε_r) and loss tangent ($tan\ \delta$) as a function of temperature were recorded using an LCR analyzer (TH2829A, Tonghui, China) attached to a programmable furnace. Samples were poled under a DC field of 6–10 kV/mm for 15 min in a silicone oil bath at 150 °C. The electrical impedance ($|Z|$) and phase angle (θ) as a function of frequency was measured using an impedance analyzer (TH2829A, Tonghui, China). The planar electromechanical coupling factor (k_p), mechanical quality factor (Q_m) and planar frequency constant (N_p) were calculated by the IEEE standard. Thermal depoling behavior was investigated by annealing the polarized samples at different temperatures for 4 h, and then the piezoelectric charge coefficient (d_{33}) was remeasured using a quasi-static d_{33} m (ZJ-6AN, IACAS, Beijing, China) when the samples were cooled to room temperature.

3. Results and Discussion

3.1. Phase Structure of Ceramics

The appearances of the oxide-modified SBT piezoceramics are presented in Figure 1. As can be seen from these figures, the pure SBT ceramic seems to be taupe; after it was doped with different oxides, different colors were presented. However, all these oxide-modified SBT piezoceramics were sintered in a uniform color and free of cracks, blotches, striations and holes, at least seen from their surfaces. The change of color also proves that the oxides as additives have dissolved into SBT, leading different color emerging mechanisms to the ceramics.

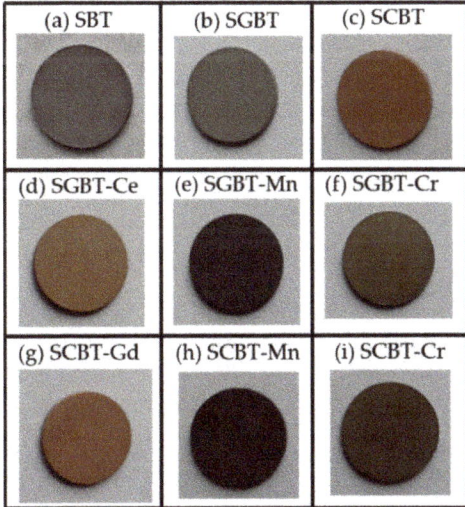

Figure 1. Appearances of the oxide-modified SBT ceramics (the corresponding chemical compositions (marked with (**a**–**i**) respectively) are located above the samples).

The XRD patterns of the oxide-modified SBT piezoceramics are shown in Figure 2. It can be seen that these samples display a single SrBi$_4$Ti$_4$O$_{15}$ phase crystallized in the orthorhombic structure with *Bb*21*m* (36) space group (JCPDS No: 43-0973). There is no impurity detected from XRD patterns, which indicates that these oxide additives have been incorporated into the crystal lattice of SrBi$_4$Ti$_4$O$_{15}$. The strongest diffraction of all these samples appears at the (1 1 9) peak, stating the fact that SrBi$_4$Ti$_4$O$_{15}$ belongs to the BLSF with the structure of four layer (m = 4) [30]. Some variations observed from the details of XRD patterns can be related with the lattice distortions of SBT caused by doping. By contrasting with the pure SBT, three diffraction peaks of the doped SBT: (0 0 10), (0 0 16) and (0 0 20) are weakened, which indicates that their grains orientating along the c-axis becomes fewer [31]. Inversely, the diffraction peaks of (2 0 0)/(0 2 0) are enhanced, which states that the number of grains oriented along the a-b plane increased.

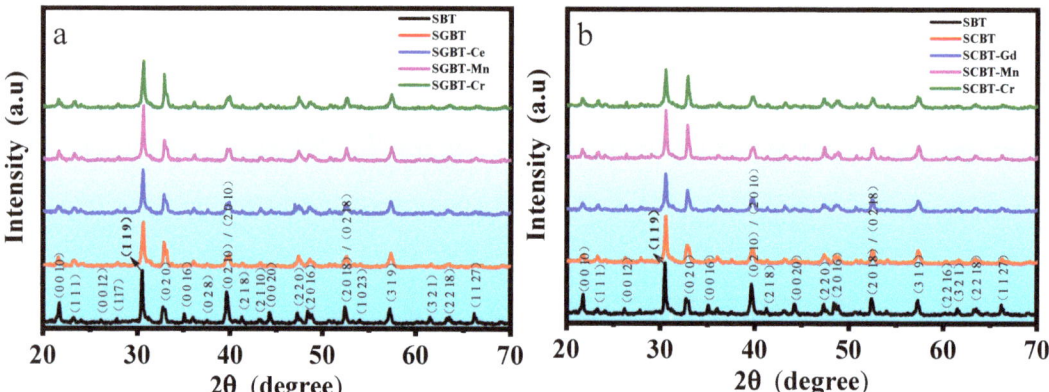

Figure 2. XRD patterns of the oxide-modified SBT ceramics: (**a**) SGBT-series; (**b**) SCBT-series.

The lattice parameters of the oxide-modified SBT ceramics are given in Table 1. The lattice parameters (a, c and v) of the oxide-modified SBT decrease, whereas the values of orthorhombic distortion (b/a) increase, which may be attributed to the ion-substitution effect caused by the addition of different oxides. The bismuth oxide layer is very strong and bismuth ion in the bismuth oxide layer are difficult to be substituted by other ions [32], meanwhile ions with similar ionic radius and same coordination number are more likely to be mutually substituted [31], consequently Sr^{2+} located at the A-site in the perovskite layers are substituted by Gd^{3+}/Ce^{4+} with smaller ion radius. Ti^{4+} located at the B-site in the perovskite layers would be substituted by Mn^{3+} and Cr^{3+}. The lattice distortion caused by ion substitution can result in the change of electrical properties for ferroelectric compounds [33]; the larger b/a value is, more distorted the lattice is. Among these compositions, the unit cell of SGBT-Cr has the largest orthorhombic distortion with a b/a value of 1.0024.

Table 1. Lattice parameters of the oxide-modified SBT ceramics.

Compositions	SBT	SGBT	SGBT-Ce	SGBT-Mn	SGBT-Cr	SCBT	SCBT-Gd	SCBT-Mn	SCBT-Cr
	Orthorhombic, $Bb2_1m$								
a (Å)	5.43177	5.42744	5.42855	5.42849	5.42483	5.42807	5.42893	5.42653	5.42411
b (Å)	5.43718	5.43507	5.43741	5.43693	5.43767	5.43818	5.43877	5.43772	5.43598
c (Å)	40.95524	40.93938	40.92796	40.88468	40.89918	40.95129	40.9501	40.95103	40.93853
V (Å^3)	1209.54	1207.65	1208.08	1206.68	1206.46	1208.83	1209.12	1208.38	1207.85
b/a	1.0010	1.0014	1.0016	1.00155	1.0024	1.0019	1.0018	1.0020	1.0022

Table 2 lists the density of the oxide-modified SBT ceramics. The relative density of SBT is measured as 94.8%, which has been changed after the addition of different oxides. According to the results given by Table 2, the addition of CeO$_2$ and MnO$_2$ played a positive effect on the densifying process of the SBT ceramic during sintering.

Table 2. Density data of the oxide-modified SBT ceramics.

Compositions	SBT	SGBT	SGBT-Ce	SGBT-Mn	SGBT-Cr	SCBT	SCBT-Gd	SCBT-Mn	SCBT-Cr
$\rho_{theoretic}$ (g/cm^3)	7.4456	7.4667	7.4614	7.4724	7.4522	7.441	7.4292	7.446	7.4367
ρ_{actual} (g/cm^3)	7.0598	6.8822	7.137	7.0692	6.3155	7.1173	7.08528	7.2209	6.5354
$\rho_{relative}$ (%)	94.8	92.2	95.7	95.7	84.7	95.6	95.4	97	87.8

3.2. Dielectric Properties of Ceramics

Figure 3 exhibits the temperature dependence of dielectric constant (ε_r) and loss tangent ($tan\ \sigma$) of the oxide-modified SBT ceramics. As can be seen, all the samples show a dielectric anomaly around 540 °C, which can be related to the ferroelectric-paraelectric phase transition of the ceramics. The peak position is considered as the Curie temperature (T_C). For the pure SBT (T_C = 537 °C, Figure 3a), a sharp rise in the values of ε_r occurred above 350 °C at low frequencies (100 Hz and 500 Hz), which can be attributed to the dielectric response of a large number of space charges to the external electric field. Moreover, its permittivity peaks are broadened and strongly dependent with frequency in terms of strength, and their positions seem to be also dependent with frequency as marked by the slightly oblique arrows. Therefore, this can be considered as a typical relaxed dielectric behavior, which is partially due to the compositional fluctuation in the crystallographic sites. In Figure 3b, SGBT exhibits a higher T_C ~ 557 °C as well as a normal phase transition. This result may be attributed to the lattice distortion of the pseduo-perovskite structure since the bivalent strontium ions were substituted by the trivalent gadolinium ions at the A-site. The tolerance factor (t) which is used for evaluating the stability of ABO$_3$-type perovskite structure can be calculated by the expression as follows [34]:

$$t = \frac{r_A + r_O}{\sqrt{2}(r_B + r_O)} \tag{1}$$

where r_A, r_B and r_O are the ionic radius of A, B and the oxygen ion, respectively. One-third of bivalent strontium ions (1.44 Å) and two-thirds of bismuth ions (1.30 Å) occupy the A site at the perovskite-like structure of pure SBT ceramics. According to the atomic percentage of the A-/B-site, the average ionic radius for Sr$_{0.92}$Gd$_{0.053}$Bi$_4$Ti$_4$O$_{15}$ could be reckoned as follows: r_A = 1/3 (0.92r_{Sr2+} + 0.053r_{Gd3+}) + 2/3 r_{Bi3+} = 1.33 Å (r_{Gd3+} = 1.107 Å), $r_B = r_{Ti4+}$ = 0.605 Å, r_{O2-} = 1.40 Å. The tolerance factor of SGBT and SBT were calculated to be 0.96 and 0.97, respectively, according to Equation (1). The reduced tolerance factor indicates that the perovskite structure of SGBT is more stable; in this case, the phase transition from the ferroelectric state to the paraelectric state needs more energy, which corresponds to a higher T_C. As can be seen from Figure 3c, T_C of SCBT (531 °C) is less low than that of SBT, which could be attributed to the reduced stability of oxygen octahedron after adding CeO$_2$ into SBT, since the coordination number of introduced Ce^{4+} is smaller than that of Sr^{2+}. The dielectric loss peak appearing around 500 °C at the low frequency of 100 Hz could be attributed to the space-charge relaxation as an extrinsic dielectric response [35]. The similar dielectric anomaly was also observed in cobalt-modified SBT [23]. The defect dipoles which are formed by combining space charges or ions with opposite charges may be slow to follow the external electric field, thereby contributing to the dielectric loss [36]. Therefore, the relaxation phenomenon reflected by the dielectric loss peaks or bumps in the wide temperature sweep can be related to the viscoelastic reorientation of defect dipoles following the external electric field at high temperature [37]. On the other hand, for all the oxide-doped compositions, the characteristic temperatures of permittivity peaks agree with that of loss peaks well. Especially, SCBT-Cr shows the most flat dielectric loss curve at 100 Hz, which indicates that the synergetic doping of CeO$_2$ and Cr$_2$O$_3$ could significantly improve the temperature stability of the dielectric properties of SBT.

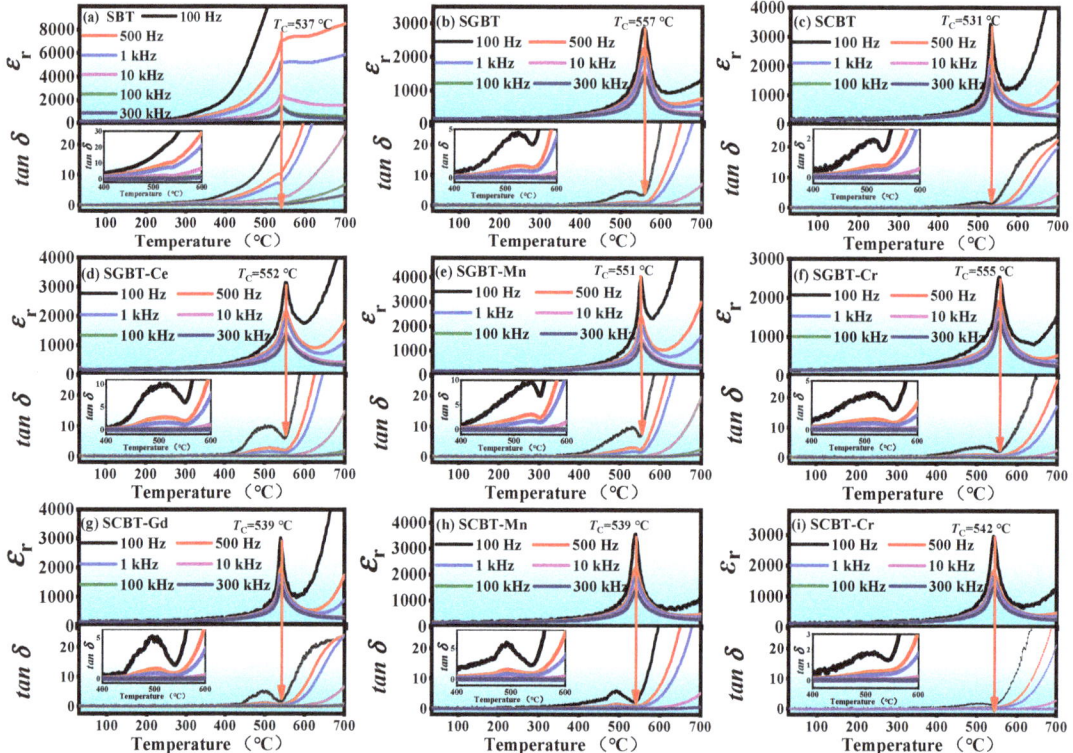

Figure 3. Temperature dependence of dielectric constant and loss tangent of the oxide-modified SBT ceramics at different frequencies.

3.3. Electro-Mechanical Coupling Property

Figure 4 shows the electro-mechanical resonance spectroscopy of the oxide-modified SBT ceramics. As can be seen, there are no resonance-antiresonance peaks in the pure SBT ceramic at the measured frequency range from 20 Hz to 2 MHz. The resonance-antiresonance peaks of SGBT and SCBT appear, respectively, at 184 kHz and 186 kHz. A high angle indicates the fully poled state of the specimen [38]. The position generated the resonance-antiresonance peak and the maximum phase angle also converts with the introduction of other additives. SCBT obtaind the maximum phase angle value ($\theta = -24.8°$), which indicates its more fully polarized degree.

Table 3 presents electro-mechanical properties of the oxide-modified SBT ceramics. Clearly, oxide additives also affect the electro-mechanical coupling properties of the SBT ceramic, especially as the addition of Cr_2O_3 has a significant impact on it. SGBT-Cr and SCBT-Cr obtain relatively high k_p, low Q_m and N_p. The oxygen vacancies in piezoceramics usually result in the increase in Q_m and the decrease in k_p for ferroelectric ceramics [39]. A higher k_p achieved by SGBT-Cr and SCBT-Cr can be attributed to the reduced oxygen vacancy concentration caused by the addition of Gd_2O_3 and CeO_2.

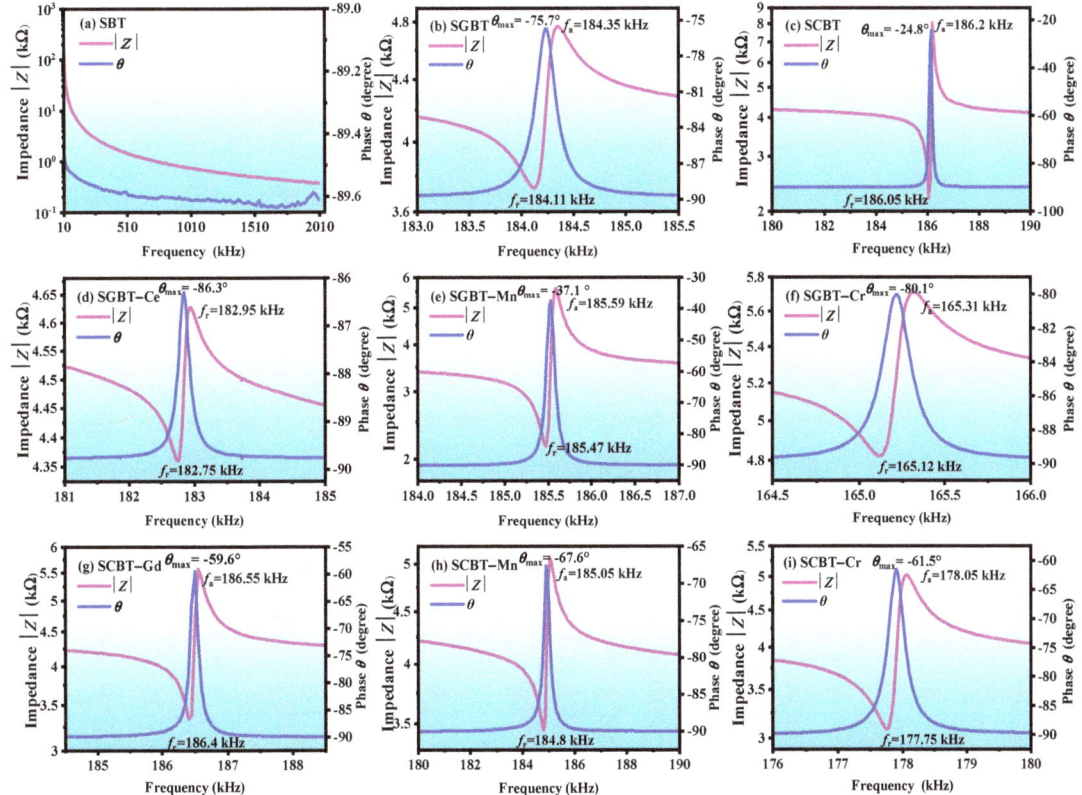

Figure 4. Electro-mechanical resonance spectroscopy of the oxide-modified SBT ceramics at room temperature.

Table 3. Electro-mechanical properties of the oxide-modified SBT ceramics.

Compositions	SGBT	SGBT-Ce	SGBT-Mn	SGBT-Cr	SCBT	SCBT-Gd	SCBT-Mn	SCBT-Cr
k_p (%)	5.72	5.2	4.03	6.35	3.7	4.5	5.8	6.51
Q_m	423	465	1240	355	1885	787	443	374
N_p (Hz·m)	2655	2745	2711	2481	2751	2740	2694	2613

3.4. Lower Limiting Frequency

Piezoelectric ceramic materials not only generate charges under the condition of applied stress or strain, but also ensure that the charges must be maintained for a period of time to be monitored by the system in actual engineering applications. The time of the maintained charge is proportional to the RC time constant. The minimum available frequency of sensor is considered to be the lower limiting frequency (f_{LL}). The relationship between RC time constant and f_{LL} is as follows:

$$f_{LL} = \frac{1}{2\pi RC} \quad (2)$$

where C is the capacitance (1 kHz) and R is the insulation resistance. Low values of f_{LL} allow the dynamic bandwidth to be extended to sonic frequencies [40]. The addition of different oxides decreases the f_{LL} of SBT as shown in the inset of Figure 5 at room temperature. The result indicates that the addition of oxides could improve the resistivity of SBT ceramics. Due to superfluous electrons generated by higher valence, Gd^{3+} and Ce^{4+} substituted lower valence Sr^{2+} can neutralize the oxygen vacancies, which increases the resistivity of SBT. The lower limiting frequency of the oxide-modified SBT ceramics at different temperatures are also compared with each other in Figure 5. The f_{LL} values of all compositions gradually increase with the rise in temperature, which may be attributed to the decrease in resistivity of the samples with increasing temperature. SCBT shows a lower f_{LL} value in the measured temperature range as compared to others. High resistivity can prevent applied electrical signals from leaking away in the process of using, only the modified SBT ceramics with high resistivity can be used in high-temperature piezoelectric fields.

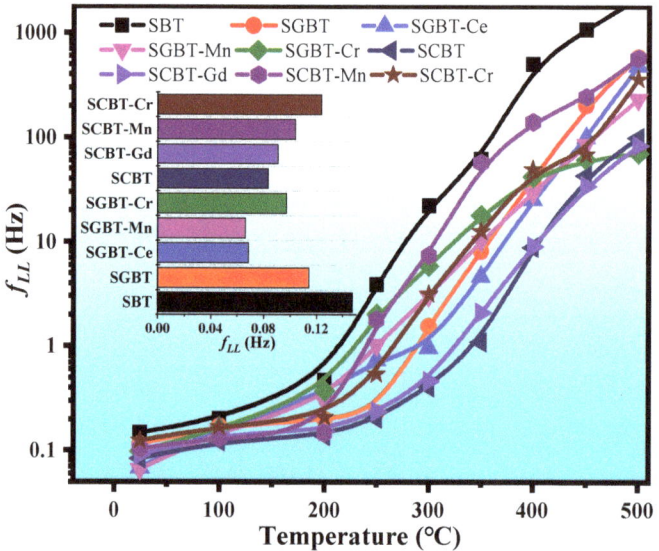

Figure 5. Lower limiting frequency of the oxide-modified SBT ceramics at different temperatures (the insert shows the f_{LL} values of various compositions at room temperature).

3.5. Piezoelectric Properties

The thermal stability of the piezoelectricity of the oxide-modified SBT ceramics is displayed in Figure 6. As can be seen from the insert, before annealing, the piezoelectric properties of the SBT ceramic (d_{33} = 10 pC/N) can be improved notably by adding only one of CeO_2 and Gd_2O_3, a higher $d_{33} \sim 22$ pC/N was achieved in SCBT and SGBT. When considering that the addition of CeO_2 and Gd_2O_3 could reduce the concentration of oxygen vacancies as mentioned above, thus the less pinning of domain walls and the elevated resistivity tend to promote the sufficient orientation of ferroelectric domains along the applied electric field during polarization. It is noteworthy that the addition of Cr_2O_3 can further enhance the piezoelectric properties of the SBT ceramic that d_{33} up to 28 pC/N was observed for SGBT-Cr and 26 pC/N for SCBT-Cr. As shown in Table 1, a larger orthorhombic distortion is obtained for SGBT-Cr and SCBT-Cr, in which a larger spontaneous polarization is believed to form [33]. Further, the thermal stability of piezoelectricity of the oxide-modified SBT ceramics was investigated by the annealing experiment. In general, the d_{33} values of all compositions slowly decrease with increasing the annealing temperature from room temperature to 400 °C, and then drastically drop after 400 °C, until they reach zero when the annealing temperature exceeded their T_C. The thermal degradation of piezoelectricity can be attributed to the decoupling of space charges at moderate temperatures and the depolarization of intrinsic dipoles at high temperatures [41]. It should be noted that the d_{33} values of SGBT-Cr were decreased by only 3.6% after being annealed at 500 °C and by 18% after being annealed at 550 °C (which is approaching T_C). This result indicates the composition with a good thermally stable piezoelectricity.

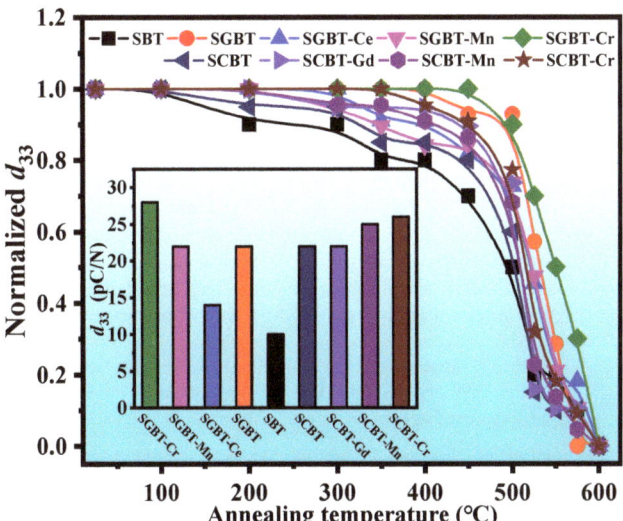

Figure 6. Thermal stability of piezoelectricity of the oxide-modified SBT ceramics (the insert shows the d_{33} values of various compositions at room temperature).

In final, dielectric and piezoelectric properties of the oxide-modified SBT ceramics were summarized in Table 4. The high piezoelectric constant, low dielectric loss, and high Curie temperature presented by some compositions demonstrated the successful modification on the SBT ceramic applied by the oxides. As compared to the modified SBT ceramics reported by other works [20,23,25], the optimized composition SGBT-Cr also possesses the competitive electrical properties with a combination of high T_C ~ 555 °C and a high d_{33} ~ 28 pC/N.

Table 4. Comparison of electrical properties between the oxide-modified SBT ceramics and other compositions reported.

Compositions	ε_r (1 kHz)	$\tan \delta$ (1 kHz)	T_c (°C)	d_{33} (pC/N)
SBT	152	0.9	537	10
SGBT	145	0.4	557	22
SGBT-Ce	168	0.4	552	14
SGBT-Mn	156	0.3	551	22
SGBT-Cr	156	0.3	555	28
SCBT	175	0.1	531	22
SCBT-Gd	169	0.2	539	22
SCBT-Mn	176	0.2	539	24
SCBT-Cr	124	0.3	542	26
SBT-Sm [20]	220	2.0	520	20
SBT-3Co [23]	200	0.6	528	28
SBT-4Mn [25]	180	0.8	530	30

4. Conclusions

The effects of oxide additives (Gd_2O_3, CeO_2, MnO_2 and Cr_2O_3) on the phase structures and electrical properties of the SBT ceramics were investigated in this work, some main results were obtained as follows: XRD patterns demonstrated that all the oxide-modified SBT ceramics were a single $SrBi_4Ti_4O_{15}$ phase. The SBT ceramics with the addition of MnO_2 presented a high relative density up to 97%. The addition of Gd_2O_3 increased the T_C of SBT, which can be related to the larger orthorhombic distortion caused by the substitution of Gd^{3+} with a smaller ionic radius for Sr^{2+} at A-site. In addition, the addition of CeO_2

reduced the T_C of SBT, based on the fact that the stability of oxygen octahedron tends to be weakened by Ce^{4+} with higher coordination number substituting for Sr^{2+} at A-site. The synergetic doping of CeO_2 and Cr_2O_3 could significantly improve the temperature stability of the dielectric properties of SBT. Cr_2O_3 can significantly enhance the k_p of SBT, at the same time, the addition of these oxides also reduced the f_{LL} of SBT at high temperatures. The addition of oxides could improve the piezoelectric property of SBT (d_{33} = 10 pC/N); in particular, SCBT-Cr and SGBT-Cr obtained a higher d_{33} of 26 pC/N and 28 pC/N, respectively. Among these compositions, SGBT-Cr ($Sr_{0.92}Gd_{0.053}Bi_4Ti_4O_{15}$ + 0.2 wt% Cr_2O_3) presented the best electrical properties, such as: T_C = 555 °C, $tan\ \delta$ = 0.4%, k_p = 6.35%, d_{33} = 28 pC/N, as well as a good thermally stable piezoelectricity that the values of d_{33} was decreased by only 3.6% after being annealed at 500 °C for 4 h and retained 82% after being annealed at the temperature approaching T_C.

Author Contributions: S.W. conceived and designed the experiments; H.Z. performed the experiments; D.W. analyzed the data; S.W. wrote the paper; L.L. and Y.C. revised the paper. All authors have read and agreed to the published version of the manuscript.

Funding: This research was funded by the Opening foundation from the Key Laboratory of Deep Earth Science and Engineering (Sichuan University), Ministry of Education (Grant number: 202007).

Institutional Review Board Statement: Not applicable.

Informed Consent Statement: Not applicable.

Data Availability Statement: The data presented in this paper can be provided at the request of the corresponding authors.

Conflicts of Interest: The authors declare no conflict of interest.

References

1. Ali, F.; Raza, W.; Li, X.; Gul, H.; Kim, K.H. Piezoelectric energy harvesters for biomedical applications. *Nano Energy* **2019**, *57*, 879–902. [CrossRef]
2. Song, K.; Zhao, R.; Wang, Z.L.; Yang, Y. Conjucted pyro-piezoelectric effect for self-powered simultaneous temperature and pressure sensing. *Adv. Mater.* **2019**, *31*, 1902831. [CrossRef] [PubMed]
3. Sahu, M.; Hajra, S.; Lee, K.; Deepti, P.L.; Mistewicz, K.; Kim, H.J. Piezoelectric nanogenerator based on lead-free flexible PVDF-barium titanate composite films for driving low power electronics. *Crystals* **2021**, *11*, 85. [CrossRef]
4. Kim, K.; Zhang, S.; Salazar, G.; Jiang, X. Design, fabrication and characterization of high temperature piezoelectric vibration sensor using YCOB crystals. *Sens. Actuators A Phys.* **2012**, *178*, 40–48. [CrossRef]
5. Parks, D.A.; Zhang, S.; Tittmann, B.R. High-temperature (>500 °C) ultrasonic transducers: An experimental comparison among three candidate piezoelectric materials. *IEEE Trans. Ultrason. Ferroelectr. Freq. Control* **2013**, *60*, 1010–1015. [CrossRef] [PubMed]
6. Shrout, T.R.; Zhang, S.J. Lead-free piezoelectric ceramics: Alternatives for PZT? *J. Electroceram.* **2007**, *19*, 113–126. [CrossRef]
7. Glaum, J.; Hoffman, M. Electric fatigue of lead-free piezoelectric materials. *J. Am. Ceram. Soc.* **2014**, *97*, 665–680. [CrossRef]
8. Vázquez-Rodríguez, M.; Jiménez, F.J.; Pardo, L.; Ochoa, P.; González, A.M.; de Frutos, J. A new prospect in road traffic energy harvesting using lead-free piezoceramics. *Materials* **2019**, *12*, 3725. [CrossRef]
9. De Araujo, A.P.; Cuchiaro, J.D.; Mcmillan, L.D.; Scott, M.C.; Scott, J.F. Fatigue-free ferroelectric capacitors with platinum electrodes. *Nature* **1995**, *374*, 627–629. [CrossRef]
10. Zhang, L.; Wang, W.; Zhou, L.; Xu, H. Bi_2WO_6 nano- and microstructures: Shape control and associated visible-light-driven photocatalytic activities. *Small* **2010**, *3*, 1618–1625. [CrossRef]
11. Wolfe, R.W.; Newnham, R.E.; Smithf, D.K.; Kay, M.I. Crystal structure of Bi_3TiNbO_9. *Ferroelectrics* **1972**, *3*, 1–7. [CrossRef]
12. Chen, Y.; Liang, D.; Wang, Q.; Zhu, J. Microstructures, dielectric, and piezoelectric properties of W/Cr co-doped $Bi_4Ti_3O_{12}$ ceramics. *J. Appl. Phys.* **2014**, *116*, 853. [CrossRef]
13. Zhang, S.T.; Sun, B.; Yang, B. $SrBi_4Ti_4O_{15}$ thin films of Ti containing bismuth-layered-ferroelectrics prepared by pulsed laser deposition. *Mater. Lett.* **2001**, *47*, 334–338. [CrossRef]
14. Ferrer, P.; Algueró, M.; Iglesias, J.E.; Castro, A. Processing and dielectric properties of $Bi_4Sr_{n-3}Ti_nO_{3n+3}$ (n = 3, 4 and 5) ceramics obtained from mechanochemically activated precursors. *J. Eur. Ceram. Soc.* **2007**, *27*, 3641–3645. [CrossRef]
15. Stevenson, T.; Martin, D.G.; Cowin, P.I.; Blumfield, A.; Bell, A.J.; Comyn, T.P.; Weaver, P.M. Piezoelectric materials for high temperature transducers and actuators. *J. Mater. Sci. Mater. Electron.* **2015**, *26*, 9256–9267. [CrossRef]
16. Nalini, G.; Row, T. Structure determination at room temperature and phase transition studies above T_c in $ABi_4Ti_4O_{15}$ (A = Ba, Sr or Pb). *Bull. Mater. Sci.* **2002**, *25*, 275–281. [CrossRef]

17. Nayak, P.; Badapanda, T.; Panigrahi, S. Dielectric, ferroelectric and conduction behavior of tungsten modified $SrBi_4Ti_4O_{15}$ ceramic. *J. Mater. Sci. Mater. Electron.* **2016**, *27*, 1217–1226. [CrossRef]
18. Ramana, E.V.; Graca, M.P.F.; Valente, M.A.; Sankaram, T.B. Improved ferroelectric and pyroelectric properties of Pb-doped $SrBi_4Ti_4O_{15}$ ceramics for high temperature applications. *J. Alloys Compd.* **2014**, *583*, 198–205. [CrossRef]
19. Rajashekhar, G.; Sreekanth, T.; James, A.R.; Ravi Kiran, U.; Sarah, P. Dielectric properties of sodium and neodymium substitute to A-Site $SrBi_4Ti_4O_{15}$ ceramics. *Ferroelectrics* **2020**, *558*, 79–91. [CrossRef]
20. Yu, L.; Hao, J.; Xu, Z.; Li, W.; Chu, R. Reddish orange-emitting and improved electrical properties of Sm_2O_3-doped $SrBi_4Ti_4O_{15}$ multifunctional ceramics. *J. Mater. Sci. Mater. Electron.* **2017**, *28*, 16341–16347. [CrossRef]
21. Nayak, P.; Badapanda, T.; Panigrahi, S. Dielectric and ferroelectric properties of Lanthanum modified $SrBi_4Ti_4O_{15}$ ceramics. *Mater. Lett.* **2016**, *172*, 32–35. [CrossRef]
22. Nayak, P.; Badapanda, T.; Singh, A.K.; Panigrahi, S. Possible relaxation and conduction mechanism in W^{6+} doped $SrBi_4Ti_4O_{15}$ ceramic. *Ceram. Int.* **2017**, *43*, 4527–4535. [CrossRef]
23. Wang, Q.; Cao, Z.P.; Wang, C.M.; Fu, Q.W.; Yin, D.F.; Tian, H.H. Thermal stabilities of electromechanical properties in cobalt-modified strontium bismuth titanate ($SrBi_4Ti_4O_{15}$). *J. Alloys Compd.* **2016**, *674*, 37–43. [CrossRef]
24. Hua, H.; Liu, H.; Ouyang, S. Structure and ferroelectric property of Nb-doped $SrBi_4Ti_4O_{15}$ ceramics. *J. Electroceram.* **2009**, *22*, 357–362.
25. Cao, Z.P.; Wang, C.M.; Lau, K.; Wang, Q.; Fu, Q.W.; Tian, H.H.; Yin, D.F. Large enhancement of piezoelectric properties in Mn-modified $SrBi_4Ti_4O_{15}$ and its thermal stabilities at elevated temperatures. *Ceram. Int.* **2016**, *42*, 11619–11625. [CrossRef]
26. Zhao, T.L.; Wang, C.M.; Wang, C.L.; Wang, Y.M.; Dong, S. Enhanced piezoelectric properties and excellent thermal stabilities of cobalt-modified Aurivillius-type calcium bismuth titanate ($CaBi_4Ti_4O_{15}$). *Mater. Sci. Eng. B* **2015**, *201*, 51–56. [CrossRef]
27. Rajashekhar, G.; Sreekanth, T.; Ravikiran, U.; Sarah, P. Dielectric properties of Na and Pr doped $SrBi_4Ti_4O_{15}$ ceramics. *Mater. Today Proc.* **2020**, *33*, 5467–5470. [CrossRef]
28. Du, H.; Ma, C.; Ma, W.; Wang, H. Microstructure evolution and dielectric properties of Ce-doped $SrBi_4Ti_4O_{15}$ ceramics synthesized via glycine-nitrate process. *Process. Appl. Ceram.* **2018**, *12*, 303–312. [CrossRef]
29. Liang, C.K.; Long, W. Microstructure and properties of Cr_2O_3-doped ternary lead zirconate titanate ceramics. *J. Am. Ceram. Soc.* **2010**, *76*, 2023–2026. [CrossRef]
30. Wang, C.M.; Wang, J.F. High performance Aurivillius phase sodium-potassium bismuth titanate lead-free piezoelectric ceramics with lithium and cerium modification. *Appl. Phys. Lett.* **2006**, *89*, 1804. [CrossRef]
31. Chen, Y.; Xie, S.; Wang, Q.; Zhu, J. Influence of Cr_2O_3 additive and sintering temperature on the structural characteristics and piezoelectric properties of $Bi_4Ti_{2.95}W_{0.05}O_{12.05}$ Aurivillius ceramics. *Prog. Nat. Sci. Mater. Int.* **2016**, *26*, 572–578. [CrossRef]
32. Newnham, R.E. Cation ordering in $Na_{0.5}Bi_{4.5}Ti_4O_{15}$. *Mater. Res. Bull.* **1967**, *2*, 1041–1044. [CrossRef]
33. Chen, Y.; Pen, Z.; Wang, Q.; Zhu, J. Crystalline structure, ferroelectric properties, and electrical conduction characteristics of W/Cr co-doped $Bi_4Ti_3O_{12}$ ceramics. *J. Alloys Compd.* **2014**, *612*, 120–125. [CrossRef]
34. Chen, Y.; Xu, J.; Xie, S.; Tan, Z.; Nie, R.; Guan, Z.; Wang, Q.; Zhu, J. Ion Doping effects on the lattice distortion and interlayer mismatch of aurivillius-type bismuth titanate compounds. *Materials* **2018**, *11*, 821. [CrossRef] [PubMed]
35. Kumar, S.; Varma, K. Structural, dielectric and ferroelectric properties of four-layer Aurivillius phase $Na_{0.5}La_{0.5}Bi_4Ti_4O_{15}$. *Mater. Sci. Eng. B* **2010**, *172*, 177–182. [CrossRef]
36. Shulman, H.S.; Damjanovic, D.; Setter, N. Niobium doping and dielectric anomalies in bismuth titanate. *J. Am. Ceram. Soc.* **2010**, *83*, 528–532. [CrossRef]
37. Chen, Y.; Xie, S.; Wang, H.; Chen, Q.; Wang, Q.; Zhu, J.; Guan, Z. Dielectric abnormality and ferroelectric asymmetry in W/Cr co-doped $Bi_4Ti_3O_{12}$ ceramics based on the effect of defect dipoles. *J. Alloys Compd.* **2017**, *696*, 746–753. [CrossRef]
38. Chen, Y.; Wang, S.; Zhou, H.; Xu, Q.; Wang, Q.; Zhu, J. A systematic analysis of the radial resonance frequency spectra of the PZT-based (Zr/Ti = 52/48) piezoceramic thin disks. *J. Adv. Ceram.* **2020**, *9*, 380–392. [CrossRef]
39. Hou, Y.D.; Lu, P.X.; Zhu, M.K.; Song, X.M.; Tang, J.L.; Wang, B.; Yan, H. Effect of Cr_2O_3 addition on the structure and electrical properties of $Pb((Zn_{1/3}Nb_{2/3})_{0.20}(Zr_{0.50}Ti_{0.50})_{0.80})O_3$ ceramics. *Mater. Sci. Eng. B* **2005**, *116*, 104–108. [CrossRef]
40. Turner, R.C.; Fuierer, P.A.; Newnham, R.E.; Shrout, T.R. Materials for high-temperature acoustic and vibration sensors—A review. *Appl. Acoust.* **1994**, *41*, 299–324. [CrossRef]
41. Chun, P. Influence of mobile space charges on the ferroelectric properties of $(K_{0.50}Na_{0.50})_2(Sr_{0.75}Ba_{0.25})_4Nb_{10}O_{30}$ ceramics. *J. Appl. Phys.* **1997**, *82*, 2528–2531.

Article

Gd/Mn Co-Doped CaBi$_4$Ti$_4$O$_{15}$ Aurivillius-Phase Ceramics: Structures, Electrical Conduction and Dielectric Relaxation Behaviors

Daowen Wu [1], Huajiang Zhou [1], Lingfeng Li [1] and Yu Chen [1,2,*]

1 School of Mechanical Engineering, Chengdu University, Chengdu 610106, China
2 Institute of Advanced Materials, Chengdu University, Chengdu 610106, China
* Correspondence: chenyu01@cdu.edu.cn; Tel.: +86-28-8461-6169

Abstract: In this work, Gd/Mn co-doped CaBi$_4$Ti$_4$O$_{15}$ Aurivillius-type ceramics with the formula of Ca$_{1-x}$Gd$_x$Bi$_4$Ti$_4$O$_{15}$ + xGd/0.2wt%MnCO$_3$ (abbreviated as CBT-xGd/0.2Mn) were prepared by the conventional solid-state reaction route. Firstly, the prepared ceramics were identified as the single CaBi$_4$Ti$_4$O$_{15}$ phase with orthorhombic symmetry and the change in lattice parameters detected from the Rietveld XRD refinement demonstrated that Gd^{3+} was successfully substituted for Ca^{2+} at the A-site. SEM observations further revealed that all samples were composed of the randomly orientated plate-like grains, and the corresponding average grain size gradually decreased with Gd content (x) increasing. For all compositions studied, the frequency independence of conductivity observed above 400 °C showed a nature of ionic conduction behavior, which was predominated by the long-range migration of oxygen vacancies. Based on the correlated barrier hopping (CBH) model, the maximum barrier height W_M, the dc conduction activation energy E_{dc}, as well as the hopping conduction activation energy E_p were calculated for the CBT-xGd/0.2Mn ceramics. The composition with x = 0.06 was found to have the highest E_{dc} value of 1.87 eV, as well as the lowest conductivity (1.8 × 10^{-5} S/m at 600 °C) among these compositions. The electrical modules analysis for this composition further illustrated the degree of interaction between charge carrier β increases, with an increase in temperature from 500 °C to 600 °C, and then a turn to decrease when the temperature exceeded 600 °C. The value of β reached a maximum of 0.967 at 600 °C, indicating that the dielectric relaxation behavior at this temperature was closer to the ideal Debye type.

Keywords: CaBi$_4$Ti$_4$O$_{15}$; ion doping; electrical conduction; dielectric relaxation; oxygen vacancies

1. Introduction

It is well-known that bismuth layer structure ferroelectrics (BLSFs) are one of the important ferroelectric oxides, which have a general formula (Bi$_2$O$_2$)$^{2+}$ (A$_{m-1}$B$_m$O$_{3m+1}$)$^{2-}$, and their crystal structure composed of pseudo-perovskite blocks (A$_{m-1}$B$_m$O$_{3m+1}$)$^{2-}$ interleaved with bismuth oxide layers (Bi$_2$O$_2$)$^{2+}$ along the c-axis [1–3]. Generally, A represents a tetravalent, pentavalent, and hexavalent ion (such as k$^+$, Li^{1+}, Zn^{2+}, Ca^{2+}, Sr^{2+}, Cr^{3+}, or La^{3+}) [4], or the mixture of them. About B, it represents a tetravalent, pentavalent, or hexavalent ion (such as Ti^{4+}, Ta^{5+}, Nd^{5+}). m is the number of BO$_6$ octahedra in the pseudo-perovskite block (m = 1, 2, 3, 4, or 5) [5]. The CaBi$_4$Ti$_4$O$_{15}$(CBT) shows the structure of A2$_1$am space group at room temperature, composing four perovskite-like TiO$_2$ octahedron units stacked in between (Bi$_2$O$_2$)$^{2+}$ layers.

For Aurivillius oxides, CBT ceramics attracted much attention from years ago, with simple preparation, transferring speed, a high fatigue strength, and low leakage current density, which are widely used in large equipment [6]. With the advancement of the aerospace industries, the research of high temperature piezoelectric acceleration sensor is urgent and necessary. Due to the high cure temperature (T_c = 790 °C) [7] and excellent

fatigue resistance [8,9], Bismuth layered piezoelectric ceramics are widely used in piezoelectric acceleration sensors. However, the low piezoelectric property limits the application of Pure CBT, because its own layer structures limit the material transportation when sintering progress and spontaneous polarization (along a-b plane) [10–12]. Moreover, a low spontaneous polarization (P_s) and higher coercive field (E_c) requires higher polarization voltage, and high electrical conductivity leads to high leakage current [13]. Therefore, it is of certain significance to study the high-temperature conductivity of CBT for operating in high-temperature environment. For Aurivillius piezoceramics, it is necessary to study electrical resistivity and conduction behavior at high temperature. Until now, many studies about CBT have been reported that concentrated on the structures and how to improve the T_c or piezoelectricity [14–16]. For example, Gd^{3+} was found to reduce the leakage current and low loss [17]. Generally, the p-type conduction is mainly a conducting type for Aurivillius piezoceramics. As such, the dc conductivity can be reduced by donor doping [18]. There are few studies about the conduction behavior of CBT. For example, Xie et al. doped W^+ into $CaBi_4Ti_4O_{15}$ piezoceramics, the relaxation activation energy of the doped system was 1.45 eV, and its hopping conduction energy was 1.50 eV, while dc conduction energy was 1.39 eV [19], but the d_{33} of this system was only 17.8 pC/N. Many studies revealed that V^{5+}, Nb^{5+}, and W^{6+} can decrease the high-temperature conductivity and increase the piezoelectric properties of BLSF ceramics, since these donor-type substituted ions could release the distortion of the oxygen octahedral, as well as reduce the concentration of oxygen vacancies in the lattice [20–22]. This means that CBT ceramics may have two different conductive types at different temperatures. However, there are many studies on the conduction mechanism of bismuth layered oxide ceramics and various mechanisms are still not widely adopted. Therefore, it would be necessary to study the conductance mechanism of the $CaBi_4Ti_4O_{15}$ ceramics, which is conducive to understanding of the microscopic motion energy of charge carriers [23].

In this work, a kind of Gd/Mn co-doped $CaBi_4Ti_4O_{15}$ ceramics were prepared using the solid-state reaction method and the structures of samples were characterized by using XRD and SEM. The effects of Gd/Mn co-doping on the electrical conduction and dielectric relaxation behaviors of $CaBi_4Ti_4O_{15}$ were studied in terms of the temperature dependent conductivity spectrum and electrical modulus analysis, with emphasis on the thermally activated motion of ionic defects, which predominates the dielectric behaviors at high temperature.

2. Experimental Section

2.1. Sample Preparation

A kind of Gd/Mn co-doped $CaBi_4Ti_4O_{15}$ piezoceramics, the formula of $Ca_{1-x}Gd_xBi_4Ti_4O_{15}$ +0.2wt%$MnCO_3$ (abbreviated as CBT-xGd/0.2Mn), were produced by the conventional solid-state reaction route. First of all, the $CaCO_3$ of 99% purity, Gd_2O_3 of 99.99% purity, TiO_2 of 99% purity, Bi_2O_3 of 99.99% purity, and $MnCO_3$ of 99% purity (as raw materials) were weighed according to the stoichiometric ratio ($CaCO_3$, Gd_2O_3, Bi_2O_3, TiO_2, and $MnCO_3$ produced in Chron Chemicals, Chengdu, China). These chemical compounds were balled for 6h with alcohol in planetary ball mill and calcined at 850 °C for 4 h in the muffle furnace. Then, the calcined powders, with 0.2 wt% $MnCO_3$, were balled for another 12 h, pressed into discs, and sintered at 1050 ~ 1150 °C for 2h to obtain the ceramic chips. Lastly, Ag paste was painted on both sides of the CBT-xGd/0.2Mn ceramics and fired at 700 °C for 10 min in air.

2.2. Sample Characterization

The phase structure for the samples were characterized through X-ray diffraction measurement (XRD, DX–2700B, Haoyuan Instrument, Dandong, China). The nature surfaces of the samples were observed using scanning electron microscope (SEM, Quanta FEG 250, FEI, Waltham, MA, USA). In the frequency range of 100~10^6 Hz, the dielectric properties were measured (room temperature ~700 °C) by an LCR meter (TH2829A, Tonghui Electronic, Changzhou, China) and the high temperature conductivity and complex impedance behavior were analyzed.

3. Results and Discussion
3.1. Phase Structures

The XRD patterns of the pure CBT and CBT-xGd/0.2Mn ceramics were performed as shown in Figure 1. It indicated the X-ray diffraction peak of the Pure CBT and the CBT-xGd/0.2Mn ceramics were consistent with the JCPDS card No.52-1640. Furthermore, all samples were orthorhombic in structure and A21am in space group. There is no other phase from the XRD pattern results of the CBT-xGd/0.2Mn ceramics, which indicated co-doping Gd/Mn formed a complete solid solution with $CaBi_4Ti_4O_{15}$; the strongest diffraction peak of the CBT-xGd/0.2Mn ceramics was (1 1 9) peak, which was consistent with the strongest diffraction peak (1 1 2m+1) of BLSF ceramics [24,25]. Compared with the pure CBT, CBT-xGd/0.2Mn ceramics showed a smaller cell volume (V) from the Table 1 of the cell paraments, and the results revealed Gd/Mn could reduce the grain size, which was valuable to increase the piezoelectricity. The amount of Gd/Mn co-doping increased had little change to the orthorhombic distortion (a/b).

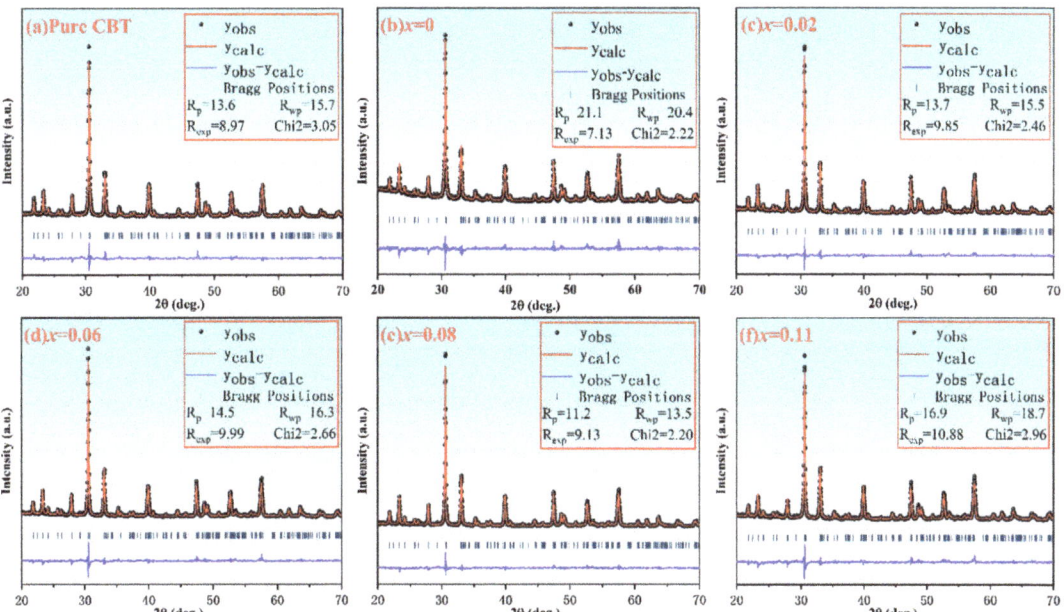

Figure 1. Rietveld analysis of XRD patterns of the CBT−xGd/0.2Mn ceramics measured at room temperature.

Table 1. Lattice parameters of the CBT-xGd/0.2Mn ceramics.

	$aBi_4Ti_4O_{15}$	CBT-xGd/0.2Mn				
		$x = 0$	$x = 0.02$	$x = 0.06$	$x = 0.08$	$x = 0.11$
a (Å)	5.431	5.428	5.426	5.428	5.426	5.426
b (Å)	5.412	5.408	5.409	5.410	5.408	5.409
c (Å)	40.73	40.71	40.71	40.74	40.73	40.74
V (Å^3)	1197.2	1194.9	1195.0	1196.2	1195	1195.5
a/b	1.0036	1.0036	1.0032	1.0034	1.0032	1.003

3.2. Microstructures

Figure 2 shows the SEM images of the CBT-xGd/0.2Mn ceramics of the original surface. It can be seen from Figure 2 that the CBT-xGd/0.2Mn ceramics presented a dense structure composed of many plate-like grains with random orientation. Such a special morphology was formed due to the structurally highly anisotropic grain growth, which had a much higher grain growth rate in the direction perpendicular to the c-axis of the BLSFs crystal [26]. Horn et al [27] reported that the (0 0 l)-type planes of the BLSFs crystal possessed a lower surface energy, which developed predominantly during sintering. Although the plate-like grains with their c-axis were normally oriented to the major surface preferred to grow up in the BLSFs ceramics, the grain orientation was random in the CBT-xGd/0.2Mn piezoceramics, as the ceramics were fabricated by pressure-less sintering.

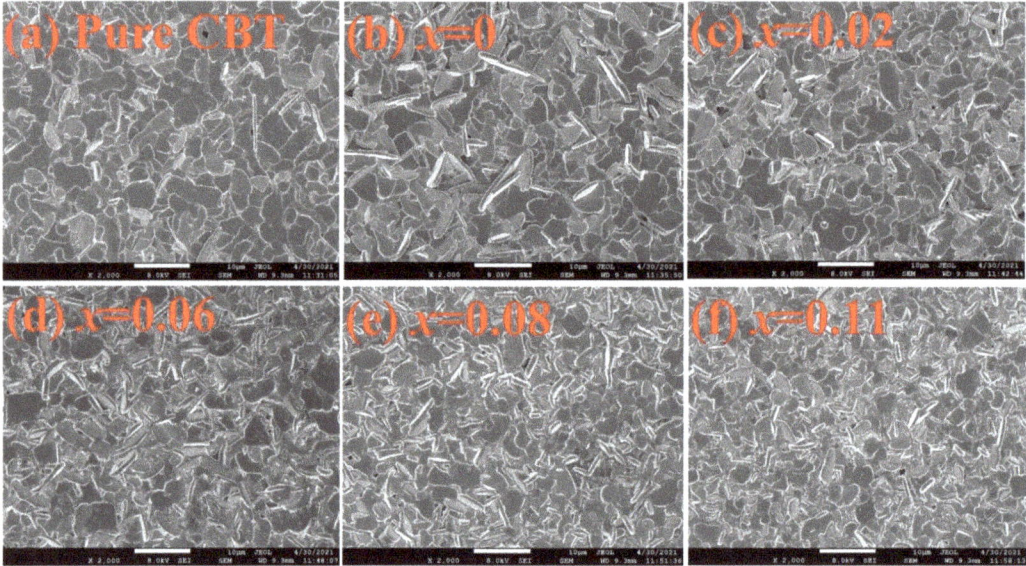

Figure 2. SEM images focused on the original surfaces of the CBT−xGd/0.2Mn ceramics.

In order to explore the grain characteristics of the CBT-xGd/0.2Mn piezoceramics quantitatively, the linear intercept method (performed by the Nano Measurer software) was used to obtain the grain size distribution from SEM images, the results were shown in Figure 3. When x increased from 0 to 0.11, the average grain size (D_λ) gradually decreased from 2.65 μm to 2.30 μm, and the corresponding size distribution became more inhomogeneous. Among the CBT-xGd/0.2Mn piezoceramics, the composition with x = 0.11 had the smallest grain size (D_λ = 2.30 μm) and the widest size distribution; such refined grains and compact structure could reduce the oxygen vacancy concentration and improve the activation energy of grain boundary, so as to provide a higher poling electric field to the ceramic. Alternatively, an obvious grain refinement, which was accompanied by more random grain orientation, occurred in those samples with $x \geqq 0.08$, indicating that enough Gd^{3+} entering into the A-site of perovskite unit would influence the growth behavior of ceramic grains. This phenomenon could be attributed to the reduced boundary energy for grain boundary migration or the increased activating energy for ion migration [28].

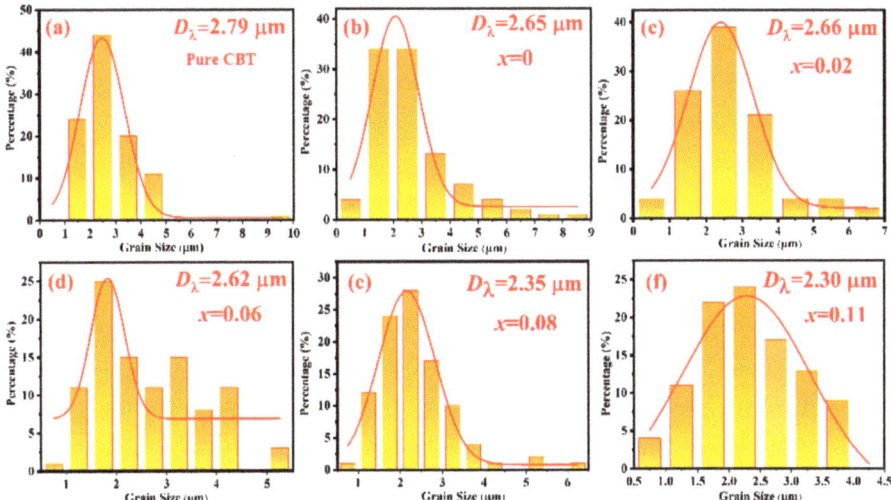

Figure 3. Grain size distribution of the CBT−xGd/0.2Mn ceramics derived from SEM images.

3.3. Electrical Conduction Behaviors

The growth of the ferroelectric phase and the movement of charge carriers are affected by conductivity to a certain extent. The study of conductivity not only helps to clarify the influence of conductivity on domain structure and its motion, but also helps to clarify the properties of carriers. The AC conductivity (σ_{ac}) of CBT-xGd/0.2Mn ceramics was studied to better understand the relaxation-conduction behaviors of the system. The AC conductivity σ_{ac} of dielectrics could be calculated using the following relation

$$\sigma_{ac} = \omega \varepsilon_0 \varepsilon_r \bullet \tan\delta \quad (1)$$

where ω is the frequency of the applied electric field, ε_0 is the permittivity of free space, ε_r is the dielectric constant, and $\tan\delta$ is the dissipation factor. Frequency dependence of σ_{ac} at various temperatures is shown in Figure 4.

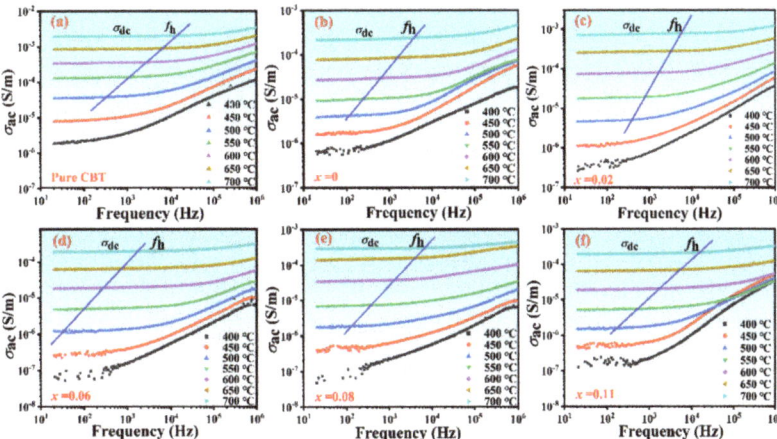

Figure 4. Frequency dependence of AC conductivity of the CBT−xGd/0.2Mn ceramics measured at different temperatures.

It can be seen from Figure 4a–f that the conductivity spectra of all samples exhibited the following characteristics: (i) The conductivity curved at lower temperatures and higher frequencies were frequency dependent, whereas at higher temperatures and lower frequencies these plots showed the frequency independence. (ii) The characteristic frequency (f_h as marked by the arrow), where the conductivity became dependent on frequency and independent on frequency, moved to a higher frequency with the temperature increasing. (iii) In the high-frequency region, the dispersion of conductivity was less and all the curves tended to merge with a single slope. The peak observed at the low frequency was due to the application of low frequency AC electric field on the high concentration doped samples, which caused the de-coupling of a large number of internal defect dipoles, resulting in a relaxed dielectric loss peak. For the perovskite-type ferroelectrics, the increase of conductivity with increasing of frequency and temperature was usually attributed to the hopping of charge carriers through the barrier or the moving of ionic defects as space charges [29,30].

The frequency dependence of conductivity has long been found to obey the following Jonscher's power law [31]:

$$\sigma_{ac} = \sigma_0 + A(T)\omega^{s(T)} \tag{2}$$

where σ_{ac} is the AC conductivity, σ_0 is the frequency independent (i. e. DC) conductivity, which can be obtained by extrapolating these plots in the low-frequency region, ω (=$2\pi f$) is the angular frequency of the AC electric field in the high-frequency region, A is a characteristic parameter assigning the polarization strength, and s is a dimensionless exponent to evaluate the degree of interaction between mobile charge carriers and surrounding lattice. Both A and the exponent s are the temperature and material intrinsic property dependent constants, which can be obtained from the fitting of the frequency dependence of conductivity according to Equation (2). The frequency and temperature dependance of ac conductivity of CBT-xGd/Mn ceramics had been carried out by Jonscher's theory, as shown in Figure 4a–f.

Ion concentrations and ion jump frequency have an influence on the conductivity of ion conductivity. The ac conductivity can be obtained by Equation (3), and the ω_p and dc conductivity are calculated by Arrhenius Equations (4) and (5), respectively [32]:

$$\omega_p = \left(\frac{\sigma_{ac}}{A}\right)^{1/s} \tag{3}$$

$$\omega_p = \omega_0 \exp(-E_h/k_B T) \tag{4}$$

$$\sigma_{dc} = \sigma_0 \exp(-E_{dc}/k_B T) \tag{5}$$

where ω_p is the hopping angular frequency, k_B is the Boltzmann constant, and T is the absolute temperature (K). Both ω_0 and σ_0 are the pre-exponential factor. E_h and E_{dc} are the activation energy of hopping conduction and dc conduction activation energy, respectively. Figure 5a shows the fitting of the σ_{ac}-f curves for the composition with x = 0.06 measured at different temperatures (500~700 °C). It can be seen that the values of s decreased with temperature increasing, indicating that the electrical conduction was a thermally activated process, which agreed with the correlated barrier hopping (CBH) model [33]. The result that s < 1 (s = back hop rare/site relaxation rate, which was defined by the jump relaxation model [34]) indicated that the time of the charge carriers returning to initial position was longer than its relaxation times. Oxygen vacancy and bismuth vacancy may cause the decrease of s value at high temperature, and the free movement of these charge carriers reduces the probability of back hoping rate.

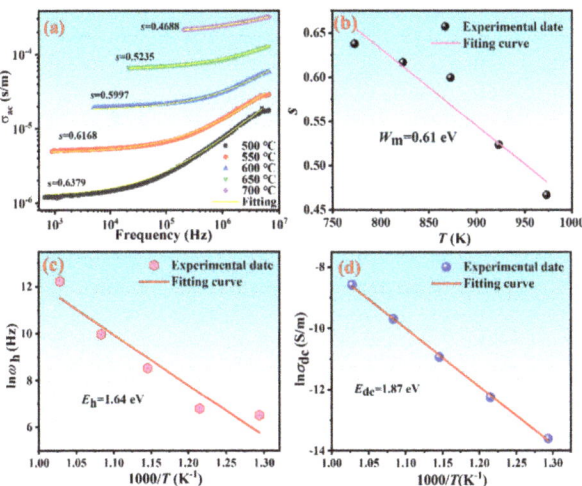

Figure 5. Fitting for the temperature and frequency dependence of conduction parameters of the composition with $x = 0.06$.

According to the CBH model, the hopping of electrons between the charged defects was limited in finite clusters, where they were bound to various defects different from the free carriers. The conduction could be attributed to the short-range hopping of localized charge carriers over trap sites separated by energy barriers of different heights. The maximum barrier height W_M, defined as the energy required to remove the electrons completely from one site to another [35], could be evaluated by using the following equation:

$$s = 1 - \frac{6k_B T}{W_M} \quad (6)$$

Figure 5b shows the fitting of the s-T curve for the composition of $x = 0.06$, according to the equation above. Before 600 °C, the obtained value of W_M (~0.61 eV) agreed well with the activation energy ($E_a = 0.3\sim0.5$ eV [36]) of single-ionized oxygen vacancies ($V_O^\bullet$), which confirmed the single-polaron hopping of electrons from the localized oxygen vacancies to the double-ionized oxygen vacancies ($V_O^\bullet \rightarrow V_O^{\bullet\bullet} + e'$) in this material. At low frequencies, electrons underwent successive and successful hopping motions for long time periods, but the ratio between successful and unsuccessful hopping, along with the relaxation of the surrounding charged carriers caused the dispersion of conductivity at high frequencies [37]. In the high-frequency region, the conductivity increase d with the increase of frequency, which may have been due to the hopping of charge carriers in finite clusters. The frequency at which the change in a slope occurs is known as the hopping frequency ω_h (=$2\pi f_h$), which obeyed the Arrhenius relation. The plots of lnf_p vs. $1000/T$ was depicted for the composition of $x = 0.06$ in Figure 5c and the value of E_h was calculated from the slope of the fitting line according to Equation (4). The E_h value is calculated to be 1.64 eV for the sample.

At low frequencies and high temperatures, the long-range migration of charge carriers contributed to the DC conductivity (σ_{dc}). With the increase of temperature, an increase in charge carrier due to thermal ionization resulted in an increased σ_{dc}. Therefore, the temperature dependence of DC conductivity could be described by the Arrhenius relation as Equation (5). Figure 5d shows the plots of $ln\sigma_{dc}$ vs. $1000/T$ and the value of E_{dc} was estimated from the fitting of the σ_{dc}-T curve based on the equation above. The fitting result estimated the value of E_{dc} to be 1.87 eV for the composition. Here, a small difference between the values of E_{dc} and E_h in the same temperature region indicated the similar type of localized charge carriers responsible for the DC and AC conduction. However, because the activation energy for the conduction process was the sum of diffusion activation (E_{dc})

and the formation energy (E_h) of charge carriers. $E_h < E_{dc}$, indicated that the hopping distance of charge carriers (usually limited in a unit-cell) was always shorter than their diffusing distance (including bulk/intragranular diffusion and grain boundary diffusion).

Table 2 listed the electrical conduction parameters of the CBT-xGd/0.2Mn ceramics calculated according to the method above. As for the pure CBT ceramic, the estimated value of DC activation energy (E_{dc} = 1.28 eV) was less than half the band gap value of CBT (E_g = 3.36 eV), indicating an extrinsic conduction process existing in the ceramic [38]. The activation energy was generally associated with the acceptor or donor levels. For the CBT-xGd/Mn ceramics, the values of W_M, E_p and E_{dc} presented a mostly consistent varying trend with the doping content of Gd. The estimated value of E_{dc} was found to increase from 1.31 eV to 1.87 eV with an increase in x from 0 to 0.06 and then showed a decrease to 1.59 eV till x = 0.11. The composition of x = 0.06, E_{dc} reached to the maximum value of 1.87 eV (associated with a relatively high E_h value of 1.64 eV), so that the composition with x = 0.06 could obtain the lowest σ_{dc} value of 1.8 × 10^{-5} S/m among the CBT-xGd/Mn ceramics.

Table 2. Electrical conduction parameters of the CBT-xGd/0.2Mn ceramics.

	CaBi$_4$Ti$_4$O$_{15}$	CBT-xGd/0.2Mn				
		x = 0	x = 0.02	x = 0.06	x = 0.08	x = 0.11
W_M (eV)	0.38	0.59	1.24	0.96	0.61	0.47
E_h (eV)	1.22	0.93	1.21	1.64	1.69	1.29
E_{dc} (eV)	1.28	1.31	1.65	1.87	1.72	1.59
σ_{dc} (S/m, 600 °C)	3.39 × 10^{-4}	2.69 × 10^{-5}	3.1 × 10^{-5}	1.8 × 10^{-5}	3.34 × 10^{-5}	1.90 × 10^{-5}

In References. [7,15], Z-Y. Shen et al. prepared a kind of Nd/Mn co-doped CBT ceramics, where the activation energy (E_a = 1.2–1.3 eV) in the temperature range 300~600 °C were suggested to be closed to the high temperature dc conductivity activation energy (E_{dc}) reported for other BLSF ceramics, which was predominated by the conduction mechanism of intrinsic charge carriers. As compared with his works, the Gd/Mn co-doped CBT ceramics prepared in this work presented a higher activation energy (E_{dc} = 1.31–1.87 eV). At high temperatures, when the intrinsic conduction predominates the material, the nominal activation was the sum of diffusion activation (E_d) and the formation energy of charge carrier (E_f). Therefore, a higher activation energy observed in our material may be owing to the different doping effect between Gd and Nd in the CBT lattice. Considering the substitution of Gd^{3+} and Nd^{3+} for Ca^{2+} at A-site, a stronger Gd–O bonds compared to Nd–O bonds might induce an increase in the formation energy of oxygen vacancies.

It was well known that the primitive BLSFs were usually not stoichiometric, since that contained amounts of inherent defects, such as oxygen vacancies and bismuth vacancies, et al. This was owing to that the unavoidable volatilization of Bi$_2$O$_3$ during the high-temperature sintering of ceramics would produce the complexes of bismuth and oxygen vacancy in the (Bi$_2$O$_2$)$^{2+}$ layers. Therefore, the doubly positively charged defects, oxygen vacancy $V_O^{\bullet\bullet}$, was considered to be the most mobile intrinsic ionic defect in the perovskite-type ferroelectrics. Their long-range migration in the octahedra of any perovskite structure, which was evidenced through greatly enhanced conductivity and activation energy of ~1 eV [39], contributed to the intrinsic ionic conduction in the temperature region of ~300 °C to ~700 °C [39].

Alternatively, according to the experimental data presented in Table 2, the variation of conductivity of the CBT-xGd/0.2Mn ceramics with the doping content of Gd (x) did not seem to be regular. The lowest conductivity value at a high temperature (600 °C) was observed at x = 0.06. For a single phase material with a homogenous microstructure, the electrical conductivity, σ, depended on both the concentration (n) and mobility (μ) of charge carriers and obeyed the following simplified equation $\sigma = nq\mu$. Here, q was the number of charges per charge carrier. With increasing x, although the oxygen vacancy concentration tended to be decreased by the donor substitution of Gd^{3+} for Ca^{2+}, the change in the mobility of oxygen vacancies could not be determined. As a result, the conductivity value at a high temperature was not expected to present a regular trend with change in x for the CBT-xGd/0.2Mn ceramics. Similar to the investigation on the doping effect in layer structured SrBi$_2$Nb$_2$O$_9$ ferroelectrics [40], the experimental results in this work suggested that the doping effects of A-site (Ca/Gd) on the dc conduction were complex, and further analysis is required to achieve a better understanding.

3.4. Electrical Impedance Analysis

To further study the dielectric relaxation behavior of the CBT-xGd/0.2Mn ceramics, Figure 6 shows the electrical modulus of x = 0.06 at a different temperature. The complex electrical modulus (M^*) was calculated from the electrical modulus measured based on the following equations:

$$M^* = M' + jM'' = j\omega C_0 Z^* \tag{7}$$

$$M' = j\omega C_0 Z'' \tag{8}$$

$$M'' = j\omega C_0 Z' \tag{9}$$

$$Z^* = Z' - jZ' \tag{10}$$

where C_0 is the capacitance of free apace given by $C_0 = \varepsilon_0 A/d$ [41], Z^* is the complex electrical impedance, and Z' and Z'' is its real part and imaginary part, respectively. It is shown that the M' values increased quickly with frequency rising at low temperature, and slowly increased gradually with frequency increasing. Moreover, the reason for the relative dispersion in low frequency region may be related to short range hopping of charge carriers and lack of recovery energy [42]. Besides, only one single peak can be seen from Figure 6b, which results from only the grain response was observed [19] at the temperature and frequencies. M'' increased sharply and reached the top may be related to both grain size and grain boundary relaxation. However, the peak value of M'' declined with the increase of temperature, which demonstrated the relaxation deviated from Debye-type relaxation. The results showed that the ions move in a hopping manner along with other related carriers [43].

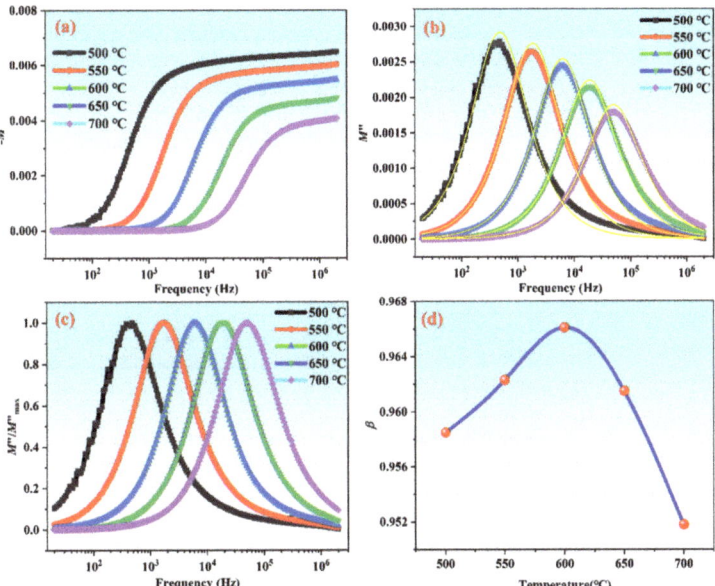

Figure 6. Electrical modulus spectroscopy of the composition with x = 0.06 measured at different temperatures: (**a**) $-M'$; (**b**) M''; (**c**) M''/M''_{max}; (**d**) temperature dependence of β.

To further clarify the dielectric relaxation mechanism, the electric modules M'' were normalized to research the relaxation process (Figure 6c). The shape of the curves were asymmetrical and higher than Debye-type relaxation. The Bergman formula [44] can explain the phenomenon:

$$M''(\omega) = \frac{M''_{max}}{1 - \beta + \frac{\beta}{1+\beta}[\beta(\omega_{max}/\omega) + (\omega/\omega_{max})]^\beta} \quad (11)$$

where M''_{max} is the maximum value of M'', ω_{max} is the angular frequency corresponding to M''_{max}, and β indicates the ideal Debye model—the closer the β value is to 1, the more it consistent with Debye-type relaxation [41]. β tended to increase to 1 as the temperature rose from 500 °C to 600 °C (Figure 6d), which showed that the relaxation type of the sample was closer to the Debye-type relaxation. A higher value of β indicated a weaker interaction between charge carriers. However, β began to decrease with increasing temperature at 600 °C, showing the dielectric relaxation behavior began to deviate from the Debye-type relaxation, which may be owing to the increased leakage current at the temperature above 600 °C. This phenomenon was consistent with that the peak value of M'' was found to decrease faster when the temperature exceeded 600 °C.

4. Conclusions

Gd/Mn co-doped $CaBi_4Ti_4O_{15}$ (CBT-xGd/0.2Mn, x = 0, 0.02, 0.06, 0.08, 0.11) ceramics were synthesized by the solid-state reaction method. The structures, electrical conduction, and dielectric relaxation behaviors of CBT-xGd/0.2Mn ceramics were studied. Some main results were obtained as follows:

(1) CBT-xGd/0.2Mn possessed a typical orthorhombic structure and the induced Gd^{3+} succeeded in substituting for Ca^{2+} at A-site. A dense microstructure composed of plate-like grains were observed in the prepared ceramics and the introduction of Gd^{3+} led to the decrease in the average grain size.
(2) The ionic conduction behavior in the high temperature region was related to the long-range migration of oxygen vacancies. The composition with x = 0.06 was found to have the highest E_{dc} value of 1.87 eV, as well as the lowest conductivity (1.8 × 10^{-5} S/m at 600 °C) among these compositions.
(3) The values of β (the degree of interaction between charge carriers) first increased and then decreased with increasing the temperature from 500 °C to 700 °C, the maximum value of 0.967 occurring at 600 °C suggested the dielectric relaxation behavior to be very close to the ideal Debye type.

Author Contributions: D.W. performed the experiment; H.Z. analyzed the experimental data; L.L. contributed reagents/materials/analysis tools; D.W. wrote the paper; Y.C. conceived and designed the experiment. All authors have read and agreed to the published version of the manuscript.

Funding: This work was supported by the State Key Laboratory of Mechanics and Control of Mechanical Structures, Nanjing University of Aeronautics and astronautics (Grant No. MCMS-E-0522G01), the State Key Laboratory of Crystal Materials, Shandong University (Grant No. KF21-08), and the Guangdong Provincial Key Laboratory of Materials and Technologies for Energy Conversion (Grant No. MATEC2022KF001).

Institutional Review Board Statement: Not applicable.

Informed Consent Statement: Not applicable.

Data Availability Statement: Not applicable.

Conflicts of Interest: The authors declare no conflict of interest.

References

1. Zeng, J.; Li, Y.; Yang, Q.; Jing, X.; Yin, Q. Grain oriented $CaBi_4Ti_4O_{15}$ piezoceramics prepared by the screen-printing multilayer grain growth technique. *J. Eur. Ceram. Soc.* **2005**, *25*, 2727–2730. [CrossRef]

2. Subbarao, E.C. A family of ferroelectric bismuth compounds. *J. Phys. Chem. Solids* **1962**, *23*, 665–676. [CrossRef]
3. Tanwar, A.; Verma, M.; Gupta, V.; Sreenivas, K. A-site substitution effect of strontium on bismuth layered $CaBi_4Ti_4O_{15}$ ceramics on electrical and piezoelectric properties. *Mater. Chem. Phys.* **2011**, *130*, 95–103. [CrossRef]
4. Montero-Tavera, C.; Durruthy-Rodríguez, M.D.; Cortés-Vega, F.D.; Yañez-Limón, J.M. Study of the structural, ferroelectric, dielectric, and pyroelectric properties of the $K_{0.5}Na_{0.5}NbO_3$ system doped with Li^+, La^{3+}, and Ti^{4+}. *J. Adv. Ceram.* **2020**, *9*, 329–338. [CrossRef]
5. Kwok, K.W.; Wong, H.Y. Piezoelectric and pyroelectric properties of Cu-doped $CaBi_4Ti_4O_{15}$ lead-free ferroelectric ceramics. *J. Phys. D: Appl. Phys.* **2009**, *42*, 095419. [CrossRef]
6. Rout, S.K.; Sinha, E.; Hussian, A.; Lee, J.S.; Ahn, C.W.; Kim, I.W.; Woo, S.I. Phase transition in A $Bi_4Ti_4O_{15}$(A= Ca, Sr, Ba) Aurivillius oxides prepared through a soft chemical route. *J. Appl. Phys.* **2009**, *105*, 024105. [CrossRef]
7. Shen, Z.Y.; Sun, H.; Tang, Y.; Li, Y.; Zhang, S. Enhanced piezoelectric properties of Nb and Mn co-doped $CaBi_4Ti_4O_{15}$ high temperature piezoceramics. *Mater. Res. Bull.* **2015**, *63*, 129–133. [CrossRef]
8. Pardo, L.; Castro, A.; Millan, P.; Alemany, C.; Jimenez, R.; Jimenez, B. $(Bi_3TiNbO_9)_x(SrBi_2Nb_2O_9)_{1-x}$ aurivillius type structure piezoelectric ceramics obtained from mechanochemically activated oxides. *Acta Mater.* **2000**, *48*, 2421–2428. [CrossRef]
9. Hong, S.H.; Trolier-McKinstry, S.; Messing, G.L. Dielectric and Electromechanical Properties of Textured Niobium—Doped Bismuth Titanate Ceramics. *J. Am. Ceram. Soc.* **2000**, *83*, 113–118. [CrossRef]
10. Cai, K.; Huang, C.; Guo, D. Significantly enhanced piezoelectricity in low-temperature sintered Aurivillius-type ceramics with ultrahigh Curie temperature of 800 °C. *J. Phys. D: Appl. Phys.* **2017**, *50*, 155302. [CrossRef]
11. Moure, A.; Castro, A.; Pardo, L. Aurivillius-type ceramics, a class of high temperature piezoelectric materials: Drawbacks, advantages and trends, Prog. *Solid State Chem.* **2009**, *37*, 15–39. [CrossRef]
12. Zhang, F.; Li, Y. Recent progress on bismuth layer-structured ferroelectrics. *J. Inorg. Mater.* **2014**, *29*, 449–460.
13. Shulman, H.S.; Testorf, M.; Damjanovic, D.; Setter, N. Microstructure, Electrical Conductivity, and Piezoelectric Properties of Bismuth Titanate. *J. Am. Ceram. Soc.* **1996**, *79*, 3124–3128. [CrossRef]
14. Yan, H.; Li, C.; Zhou, J.; Zhu, W.; He, L.; Song, Y.; Yu, Y. Effects of A-site (NaCe) substitution with Na-deficiency on structures and properties of $CaBi_4Ti_4O_{15}$-based high-Curie-temperature ceramics. *Jpn. J. Appl. Phys.* **2001**, *40*, 6501. [CrossRef]
15. Shen, Z.Y.; Luo, W.Q.; Tang, Y.; Zhang, S.; Li, Y. Microstructure and electrical properties of Nb and Mn co-doped $CaBi_4Ti_4O_{15}$ high temperature piezoceramics obtained by two-step sintering. *Ceram. Int.* **2016**, *42*, 7868–7872. [CrossRef]
16. Peng, Z.; Huang, F.; Chen, Q.; Bao, S.; Wang, X.; Xiao, D.; Zhu, J. Microstructure and impedance analysis of $CaBi_4Ti_4O_{15}$ piezoceramics with (LiCe)-modifications. In Proceedings of the ISAF-ECAPD-PFM 2012, Aveiro, Portugal, 9–13 July 2012; pp. 1–4.
17. Groń, T.; Maciejkowicz, M.; Tomaszewicz, E.; Guzik, M.; Oboz, M.; Sawicki, B.; Pawlus, S.; Nowok, A.; Kukuła, Z. Combustion synthesis, structural, magnetic and dielectric properties of Gd^{3+}-doped lead molybdato-tungstates. *J. Adv. Ceram.* **2020**, *9*, 255–268. [CrossRef]
18. Long, C.; Fan, H.; Li, M. High temperature Aurivillius piezoelectrics: The effect of (Li, Ln) modification on the structure and properties of $(Li,Ln)_{0.06}(Na,Bi)_{0.44}Bi_2Nb_2O_9$ (Ln= Ce, Nd, La and Y). *Dalton Trans.* **2013**, *42*, 3561–3570. [CrossRef]
19. Xie, X.; Zhou, Z.; Wang, T.; Liang, R.; Dong, X. High temperature impedance properties and conduction mechanism of W^{6+}-doped $CaBi_4Ti_4O_{15}$ Aurivillius piezoceramics. *J. Appl. Phys.* **2018**, *124*, 204101. [CrossRef]
20. Xie, X.; Zhou, Z.; Gao, B.; Zhou, Z.; Liang, R.; Dong, X. Ion-Pair Engineering-Induced High Piezoelectricity in $Bi4Ti3O12$-Based High-Temperature Piezoceramics. *ACS Appl. Mater. Interfaces* **2022**, *14*, 14321–14330. [CrossRef]
21. Zhou, Z.; Dong, X.; Chen, H.; Yan, H. Structural and electrical properties of W^{6+}-doped Bi_3TiNbO_9 high-temperature piezoceramics. *J. Am. Ceram. Soc.* **2006**, *89*, 1756–1760. [CrossRef]
22. He, X.; Wang, B.; Fu, X.; Chen, Z. Structural, electrical and piezoelectric properties of V-, Nb-and W-substituted $CaBi_4Ti_4O_{15}$ ceramics. *J. Mater. Sci. Mater. Electron.* **2014**, *25*, 3396–3402. [CrossRef]
23. Badapanda, T.; Harichandan, R.; Kumar, T.B.; Mishra, S.R.; Anwar, S. Dielectric relaxation and conduction mechanism of dysprosium doped barium bismuth titanate Aurivillius ceramics. *J. Mater. Sci. Mater. Electron.* **2017**, *28*, 2775–2787. [CrossRef]
24. Peng, D.; Wang, X.; Xu, C.; Yao, X.; Lin, J.; Sun, T. Bright upconversion emission, increased Tc, enhanced ferroelectric and piezoelectric properties in Er-doped $CaBi_4Ti_4O_{15}$ multifunctional ferroelectric oxides. *J. Am. Ceram. Soc.* **2013**, *96*, 184–190. [CrossRef]
25. Yan, H.; Li, C.; Zhou, J.; Zhu, W.; He, L.; Song, Y. A-site (MCe) substitution effects on the structures and properties of $CaBi_4Ti_4O_{15}$ ceramics. *Jpn. J. Appl. Phys.* **2000**, *39*, 6339. [CrossRef]
26. Du, H.; Shi, X.; Li, H. Phase developments and dielectric responses of barium substituted four-layer $CaBi_4Ti_4O_{15}$ Aurivillius. *Bull. Mater. Sci.* **2011**, *34*, 1201–1207. [CrossRef]
27. Horn, J.A.; Zhang, S.C.; Selvaraj, U.; Messing, G.L.; Trolier-McKinstry, S. Templated grain growth of textured bismuth titanate. *J. Am. Ceram. Soc.* **1999**, *82*, 921–926. [CrossRef]
28. Hou, J.; Kumar, R.V.; Qu, Y.; Krsmanovic, D. B-site doping effect on electrical properties of $Bi_4Ti_{3-2x}Nb_xTa_xO_{12}$ ceramics. *Scr. Mater.* **2009**, *61*, 664–667. [CrossRef]
29. Bidault, O.; Goux, P.; Kchikech, M.; Belkaoumi, M.; Maglione, M. Space-charge relaxation in perovskites. *Phys. Rev. B* **1994**, *49*, 7868. [CrossRef]
30. Scott, J.F.; Dawber, M. Oxygen-vacancy ordering as a fatigue mechanism in perovskite ferroelectrics. *Appl. Phys. Lett.* **2000**, *76*, 3801–3803. [CrossRef]

31. Jonscher, A.K. The 'universal' dielectric response. *Nature* **1977**, *267*, 673–679. [CrossRef]
32. Almond, D.P.; Duncan, G.K.; West, A.R. The determination of hopping rates and carrier concentrations in ionic conductors by a new analysis of ac conductivity. *Solid State Ion.* **1983**, *8*, 159–164. [CrossRef]
33. Pike, G.E. AC conductivity of scandium oxide and a new hopping model for conductivity. *Phys. Rev. B* **1972**, *6*, 1572. [CrossRef]
34. Funke, K.; Roling, B.; Lange, M. Dynamics of mobile ions in crystals, glasses and melts. *Solid State Ion.* **1998**, *105*, 195–208. [CrossRef]
35. Chaudhuri, B.K.; Chaudhuri, K.; Som, K.K. Concentration dependences of DC conductivity of the iron-bismuth oxide glasses—I. *J. Phys. Chem. Solids* **1989**, *50*, 1137–1147. [CrossRef]
36. Liu, L.; Huang, Y.; Su, C.; Fang, L.; Wu, M.; Hu, C.; Fan, H. Space-charge relaxation and electrical conduction in $K_{0.5}Na_{0.5}NbO_3$ at high temperatures. *Appl. Phys. A* **2011**, *104*, 1047–1051. [CrossRef]
37. Nayak, P.; Badapanda, T.; Singh, A.K.; Panigrahi, S. Possible relaxation and conduction mechanism in W^{6+} doped $SrBi_4Ti_4O_{15}$ ceramic. *Ceram. Int.* **2017**, *43*, 4527–4535. [CrossRef]
38. Tanwar, A.; Sreenivas, K.; Gupta, V. Effect of orthorhombic distortion on dielectric and piezoelectric properties of CaBi4Ti4O15 ceramics. *J. Appl. Phys.* **2009**, *105*, 084105. [CrossRef]
39. Macedo, Z.S.; Ferrari, C.R.; Hernandes, A.C. Impedance spectroscopy of $Bi_4Ti_3O_{12}$ ceramic produced by self-propagating high-temperature synthesis technique. *J. Eur. Ceram. Soc.* **2004**, *24*, 2567–2574. [CrossRef]
40. Wu, Y.; Forbess, M.J.; Seraji, S.; Limmer, S.J.; Chou, T.P.; Nguyen, C.; Cao, G. Doping effect in layer structured SrBi2Nb2O9 ferroelectrics. *J. Appl. Phys.* **2001**, *90*, 5296–5302. [CrossRef]
41. Mahamoud, H.; Louati, B.; Hlel, F.; Guidara, K. Impedance and modulus analysis of the $(Na_{0.6}Ag_{0.4})_2PbP_2O_7$ compound. *J. Alloys Compd.* **2011**, *509*, 6083–6089. [CrossRef]
42. Mandal, S.K.; Dey, P.; Nath, T.K. Structural, electrical and dielectric properties of $La_{0.7}Sr_{0.3}MnO_3$–$ErMnO_3$ multiferroic composites. *Mater. Sci. Eng. B* **2014**, *181*, 70–76. [CrossRef]
43. Mohanty, V.; Cheruku, R.; Vijayan, L.; Govindaraj, G. Ce-substituted lithium ferrite: Preparation and electrical relaxation studies. *J. Mater. Sci. Technol.* **2014**, *30*, 335–341. [CrossRef]
44. Zurbuchen, M.A.; Sherman, V.O.; Tagantsev, A.K.; Schubert, J.; Hawley, M.E.; Fong, D.D.; Streiffer, S.K.; Jia, Y.; Tian, W.; Schlom, G. Synthesis, structure, and electrical behavior of $Sr_4Bi_4Ti_7O_{24}$. *J. Appl. Phys.* **2010**, *107*, 024106. [CrossRef]

Article

Microstructure and Mechanical Properties of Composites Obtained by Spark Plasma Sintering of Ti$_3$SiC$_2$-15 vol.%Cu Mixtures

Rui Zhang [1,2,3,*], Biao Chen [4], Fuyan Liu [5,*], Miao Sun [1], Huiming Zhang [1] and Chenlong Wu [4]

[1] School of Mechanical Engineering, Chengdu University, Chengdu 610106, China; 17852583676@163.com (M.S.); zhanghuiming@stu.cdu.edu.cn (H.Z.)
[2] State Key Laboratory of Solid Lubrication, Lanzhou Institute of Chemical Physics, Chinese Academy of Sciences, Lanzhou 730000, China
[3] Sichuan Province Engineering Technology Research Center of Powder Metallurgy, Chengdu University, Chengdu 610106, China
[4] School of Mechanical Engineering, Xinjiang University, Urumqi 830000, China; chenbiao@stu.xju.edu.cn (B.C.); w996940130@163.com (C.W.)
[5] School of Chemical Engineering and Materials, Changzhou Institute of Technology, Changzhou 213032, China
* Correspondence: zhangrui0214@cdu.edu.cn (R.Z.); liufy@czu.cn (F.L.)

Citation: Zhang, R.; Chen, B.; Liu, F.; Sun, M.; Zhang, H.; Wu, C. Microstructure and Mechanical Properties of Composites Obtained by Spark Plasma Sintering of Ti$_3$SiC$_2$-15 vol.%Cu Mixtures. *Materials* **2022**, *15*, 2515. https://doi.org/10.3390/ma15072515

Academic Editors: Mattia Biesuz and Dina Dudina

Received: 13 February 2022
Accepted: 28 March 2022
Published: 29 March 2022

Publisher's Note: MDPI stays neutral with regard to jurisdictional claims in published maps and institutional affiliations.

Copyright: © 2022 by the authors. Licensee MDPI, Basel, Switzerland. This article is an open access article distributed under the terms and conditions of the Creative Commons Attribution (CC BY) license (https://creativecommons.org/licenses/by/4.0/).

Abstract: Method of soft metal (Cu) strengthening of Ti$_3$SiC$_2$ was conducted to increase the hardness and improve the wear resistance of Ti$_3$SiC$_2$. Ti$_3$SiC$_2$/Cu composites containing 15 vol.% Cu were fabricated by Spark Plasma Sintering (SPS) in a vacuum. The effect of the sintering temperature on the phase composition, microstructure and mechanical properties of the composites was investigated in detail. The as-synthesized composites were thoroughly characterized by scanning electron micrography (SEM), optical micrography (OM) and X-ray diffractometry (XRD), respectively. The results indicated that the constituent of the Ti$_3$SiC$_2$/Cu composites sintered at different temperatures included Ti$_3$SiC$_2$, Cu$_3$Si and TiC. The formation of Cu$_3$Si and TiC originated from the reaction between Ti$_3$SiC$_2$ and Cu, which was induced by the presence of Cu and the de-intercalation of Si atoms Ti$_3$SiC$_2$. OM analysis showed that with the increase in the sintering temperature, the reaction between Ti$_3$SiC$_2$ and Cu was severe, leading to the Ti$_3$SiC$_2$ getting smaller and smaller. SEM measurements illustrated that the uniformity of the microstructure distribution of the composites was restricted by the agglomeration of Cu, controlling the mechanical behaviors of the composites. At 1000 °C, the distribution of Cu in the composites was relatively even; thus, the composites exhibited the highest density, relatively high hardness and compressive strength. The relationships of the temperature, the current and the axial dimension with the time during the sintering process were further discussed. Additionally, a schematic illustration was proposed to explain the related sintering characteristic of the composites sintered by SPS. The as-synthesized Ti$_3$SiC$_2$/Cu composites were expected to improve the wear resistance of polycrystalline Ti$_3$SiC$_2$.

Keywords: MAX phase; SPS; Ti$_3$SiC$_2$/Cu composites; sintering characteristic

1. Introduction

As a typical ternary layered MAX phase, Ti$_3$SiC$_2$ exhibited combined characteristics of both metals and ceramics. It possessed good machinability, good electrical and thermal conductivity, good ductility, thermal stability, oxidation resistance, etc. [1–4]. It has potential for applications in electro friction materials, novel structural/functional ceramic materials and high-temperature lubricating materials.

The crystalline structure of Ti$_3$SiC$_2$ can be described as a sandwich structure: a layer of Si and the twin boundary of TiC. The chemical bonding between the Si atom and the other atoms, such as Ti and C, was to a certain extent relatively weak compared to the strong Ti-C bonding [5,6]. Based on its unique structure, Ti$_3$SiC$_2$ was chemically reactive when it contacted with metals at high temperatures. Li et al. [7] fabricated Ti$_3$SiC$_2$/Ni and Ti_3SiC_2/Co through vacuum sintering. It was found that the metals (Ni and Co) were prone to aggregating towards their surfaces due to the poor wettability between Ti_3SiC_2 and Ni or Co. Gu et al. [8] investigated the possibility of fabricating Ti_3SiC_2-Ti composites. They discovered that the as-obtained composites were mainly composed of Ti_3SiC_2, TiC_x, Ti_5Si_3, and $TiSi_2$. It was inferred that the decomposition of Ti_3SiC_2 was owing to the de-intercalation of Si and the separation of carbon from Ti_3SiC_2. Gu et al. [9] studied the reactions between Ti_3SiC_2 and Al in the temperature range of 600–650 °C, and they found that besides Ti_3SiC_2 and Al, new phases Al_3Ti, Al_4SiC_4 and Al_4C_3 were generated, the formation of which relied on the time, temperature and the relative amount of Al and Ti_3SiC_2. Kothalkar et al. [10] synthesized NiTi-Ti_3SiC_2 composite, and they discovered that the composites showed higher damping up to applied stress of 200 MPa. For example, the energy dissipation of the composite was thirteen times larger than pure Ti_3SiC_2 and two times larger than pure NiTi.

Except for the Ti_3SiC_2-metals composites described above, researchers concentrated on the investigation of Ti_3SiC_2/Cu composites. Dang et al. [11] fabricated Ti_3SiC_2/Cu composites with different contents of Cu by mechanical alloying and spark plasma sintering. They inferred that the presence of Cu leads to the decomposition of Ti_3SiC_2 to form TiC_x, $Ti_5Si_3C_y$, Cu_3Si, and $TiSi_2C_z$. In another paper, Dang et al. [12] synthesized Ti_3SiC_2/Cu/Al/SiC composites by powder metallurgy/spark plasma sintering and found that the addition of Al could inhibit the decomposition of Ti_3SiC_2. Lu et al. [5] found chemical reaction between Cu and Ti_3SiC_2 contributes to the wettability. Zhou and his colleagues [13] investigated the chemical reactions and stability of Ti_3SiC_2 in Cu for the Cu/Ti_3SiC_2 composites. They found that at low content of Ti_3SiC_2 or below 1000 °C, Cu (Si) solid solution and TiC_x were generated, whereas at high temperature or high content of Ti_3SiC_2, Cu-Si intermetallic compounds, such as Cu_5Si, $Cu_{15}Si_4$, and TiC_x, were generated. In our recent publication [14], Ti_3SiC_2/Cu composites were synthesized by spark plasma sintering technique at various temperatures. The microstructure, composition and mechanical properties of the as-obtained composites were investigated. The results indicated that the Ti_3SiC_2/Cu composite sintered at 1100 °C exhibited superior mechanical properties. While these results give partial information on the reactions occurring between Ti_3SiC_2 and Cu, they did not permit elucidation of the entire sintering behaviors occurring during the sintering process.

In this paper, the Ti_3SiC_2/Cu composites were fabricated by Spark Plasma Sintering (SPS) at different sintering temperatures. The sintering behaviors of the composites were explored by the characterization of the phase composition, microstructure and mechanical properties of the composites based on the relationship of the temperature, the current and the pressure with the time during the sintering process.

2. Experiment

2.1. Samples Preparation

Ti_3SiC_2/Cu composites were fabricated using powder mixture of Ti_3SiC_2 (average particle size: 38 μm, ≥98% purity, 11 technology Co., Ltd., Jilin, China) and Cu (average particle size: 74 μm, ≥99.9% purity, Macklin Biochemical Co., Ltd., Shanghai, China). The content of Cu in Ti_3SiC_2/Cu composites was 15 vol.%. The mixture with designed composition was first mixed by a ball-milling machine (PMQD2LB, Nanjing Chishun Technology Development Co., Ltd., Nanjing, China) with a rotational speed of 150 rpm and a ratio of ball to powder of 3 for 6 h, then filled into a graphite die (inner diameter: Φ25 mm). Finally, it was sintered by Spark Plasma Sintering (SPS, Model Labox-350, Xinxie, Japan) at different sintering temperatures (950, 1000 and 1050 °C, respectively) under a pressure of 35 MPa in vacuum for 20 min and cooled with the furnace. The heating rate was set as follows. From the room temperature to 600 °C, the heating rate was 100 °C/min. Additionally, from 600 °C to the desired sintering temperature, the heating rate was 50 °C/min.

2.2. Mechanical Property

The density of the as-prepared Ti_3SiC_2/Cu composites was measured by the Archimedes method. Cylinder specimens with a size of $\varphi 5$ mm $\times$ 12 mm were machined for compressive strength testing, which was performed on WDW-100 universal materials testing machine (Jinan Hansen Precision Instrument Co., Ltd., Jinan, China). The Vickers hardness of the composites was determined in an MHVD-50AP microhardness tester (Shanghai Jujing Precision Instrument Manufacturing Co., Ltd., Shanghai, China) at a load of 1 kg with a dwell time of 10 s.

2.3. Analysis

The as-synthesized Ti_3SiC_2/Cu composites were polished using 0.5 um polishing paste by an automatic polishing machine (AutoMetTM250, Yigong Testing and Measuring Instrument Co., Ltd. Shanghai, China) for microscopic evaluation. In order to expose the grains, the polished samples were etched using a 1:1:1 by volume $HF:HNO_3:H_2O$ solution and observed under optical microscopy (MDJ-DM, Chongqing Auto Optical Instrument Co., Ltd., Chongqing, China). The compression fracture morphology of Ti_3SiC_2/Cu composites was observed by scanning electron microscopy (SEM, JSM-6510LA, JEOL Japan Electronics Co., Ltd., Zhaodao, Japan) equipped with energy dispersive spectroscopy (EDS). X-ray diffraction (XRD) analysis was carried out on a DX-2700B diffractometer (Dandong Haoyuan Instrument Co., Ltd., Dandong, China) with Cu Kα radiation at a scanning rate of 7.2 °/min to identify the phase composition of Ti_3SiC_2/Cu composites.

3. Results and Discussion

3.1. Phase Composition and Microstructure

Figure 1 shows XRD patterns of the as-synthesized Ti_3SiC_2/Cu composites at different sintered temperatures. It was clearly observed that the as-synthesized Ti_3SiC_2/Cu composites were all composed of Ti_3SiC_2, Cu_3Si and TiC, which was independent of the sintered temperature.

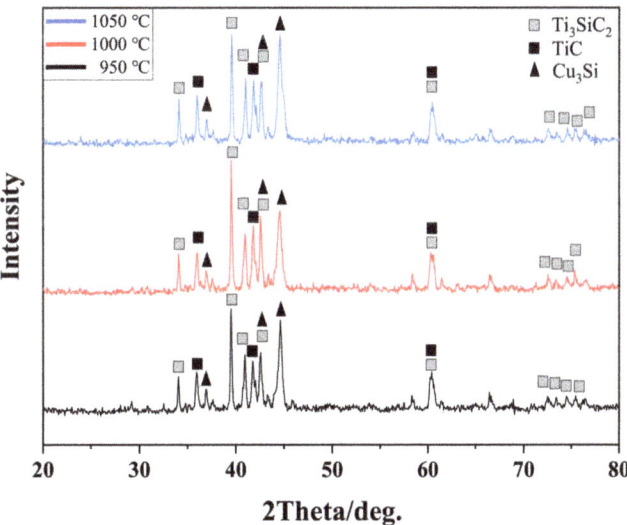

Figure 1. XRD patterns of Ti_3SiC_2/Cu composites at different sintered temperatures.

Based on the unique sandwich structure of Ti_3SiC_2, it exhibited high reactivity when contacting with metal phases [5,11,15–19]. On the one hand, the Si atoms were easily deintercalated from Ti_3SiC_2. On the other hand, Si-containing solid solution or intermetallic compounds was prone to form when the metal phases contacted with Ti_3SiC_2. The presence of contacted metal phases accelerated the decomposition of Ti_3SiC_2. As for Ti_3SiC_2/Cu composites, Cu can form Cu (Si) solid solution or react with Si to form Cu_xSi_y intermetallic compounds [18]. Therefore, the composition of Ti_3SiC_2/Cu composites consisted of Cu(Si) solid solution, Cu_xSi_y intermetallic compounds, and TiC_z, which were commonly examined by researchers [5,11,13].

In this study, with the change in sintered temperatures, Cu_3Si was the only Cu-Si intermetallic compound, and TiC was the only decomposed product of Ti_3SiC_2. According to Guo et al. [20], when the content of Cu was less than that of Si, Cu_3Si was the preferential product of the Cu-Si system. Because based on the binary phase diagram of Cu-Si, the content of Si in Cu_3Si (22.2–25.2%) was the maximum among six copper silicides (Cu (Si), Cu_7Si, Cu_5Si, Cu_4Si, $Cu_{15}Si_4$ and Cu_3Si) [20]. This explained the reason why Cu_3Si was the single Cu-Si intermetallic compound in our present study.

The effect of sintered temperature on the microstructure is shown in Figure 2. As seen in Figure 2a, at 950 °C, the typical plate-like morphology of Ti_3SiC_2 grains was evidently inhibited by the addition of Cu. A considerable amount of Ti_3SiC_2 equiaxed grains appeared due to the reaction between Ti_3SiC_2 and Cu. Additionally, with the increase in the sintered temperature from 950 °C to 1050 °C, the Ti_3SiC_2 granules decreased, which was accompanied by fewer pores or holes, indicating higher reactivity of Ti_3SiC_2 and Cu.

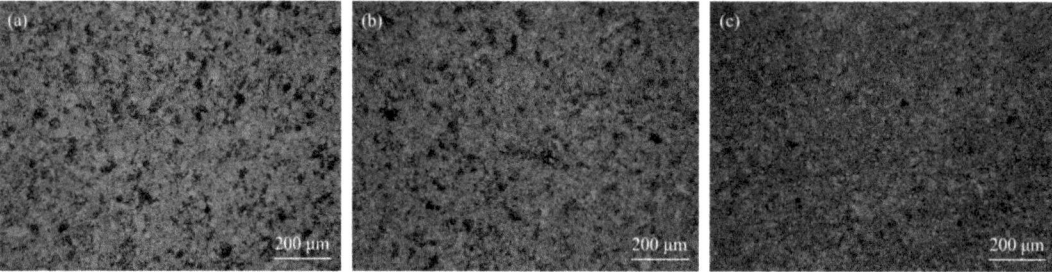

Figure 2. Optical micrographs of polished and etched Ti_3SiC_2/Cu composites sintered at 950 °C (**a**), 1000 °C (**b**) and 1050 °C (**c**).

The back scattering electron images of polished and etched Ti_3SiC_2/Cu composites sintered at 950 °C, 1000 °C and 1050 °C are shown in Figure 3. As mentioned above, the main composition of Ti_3SiC_2/Cu composites were Ti_3SiC_2, Cu_3Si and TiC. Ti_3SiC_2 was located at the dark grey area in Figure 3; both Cu_3Si and TiC were located at the light grey area in Figure 3. It was clearly seen from Figure 3 that the as-formed Cu_3Si distributed along the grain boundary of Ti_3SiC_2, and it was accompanied by the formation of hard TiC particles. Moreover, with the increase in the sintered temperature, the reaction between Ti_3SiC_2 and Cu became more severe, resulting in the formation of Cu_3Si with a non-negligible amount, especially at 1000 °C. The reaction mechanism between Ti_3SiC_2 and Cu was proposed as follows. After milling, the Cu powder is relatively evenly distributed in Ti_3SiC_2 powder. At elevated temperatures (for example, 900 °C), the Si atoms de-intercalated from Ti_3SiC_2 grains, diffused rapidly around Cu and reacted with Cu to form Cu_3Si. Meanwhile, the original Ti_3SiC_2 skeleton structure transformed to TiC structure, which was concomitant with the formation of pores or holes due to the mismatch of skeletal density between Ti_3SiC_2 and TiC. When the sintered temperature raised (for example, 1000 °C), the de-intercalation of Si atoms from the Ti_3SiC_2 skeleton was accelerated by the defects (pores or holes) and high temperature. On the other hand, Cu had a tendency to melt and flow around the grain boundaries of Ti_3SiC_2 grains and the as-formed defects mentioned above. All these led to more violent reactions between Ti_3SiC_2 and Cu, causing the Ti_3SiC_2 grains to become smaller and smaller. This reaction process well coincided with

Zhou et al. [13]. Theoretically, with the increase in temperature, the reactivity of Ti$_3$SiC$_2$ and Cu increased, producing more Cu$_3$Si. However, when the sintered temperature reached 1050 °C, a large amount of Cu melted and released from the graphite die, leading to the loss of Cu. As seen in Figure 3, visually, the content of Cu$_3$Si was largest at 1000 °C, not at 1050 °C.

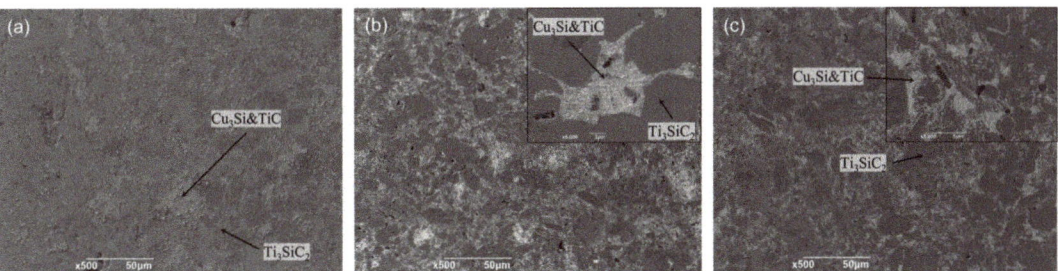

Figure 3. BSE micrographs of the Ti$_3$SiC$_2$/Cu composites sintered at (**a**) 950 °C, (**b**) 1000 °C (inset showed the structure at a higher magnification) and (**c**) 1050 °C (inset showed the structure at a higher magnification).

Figure 4 illustrates the mapping of elemental distribution for Ti$_3$SiC$_2$/Cu composites sintered at different temperatures. It was irrefutable evidence for the explanation of the sintering effect of the composites. At 950 °C, the Cu considerably aggregated in the Ti$_3$SiC$_2$/Cu composites (see Figure 4a). At this temperature, a solid–solid sintering process occurred. The agglomeration of Cu relied on the uniformity of its distribution in Ti$_3$SiC$_2$ during ball milling. As we know, ball milling cannot avoid material agglomeration. At 1000 °C, it was obviously observed that Cu was relatively evenly distributed in the Ti$_3$SiC$_2$/Cu composites (see Figure 4b). It was speculated that solid–liquid sintering process took place at 1000 °C. The appropriate flowability of the quasi-liquid Cu contributed to not only its reaction with Si atoms but also its uniform distribution in the composites. At 1050 °C, the Cu melted and rapidly flowed around the Ti$_3$SiC$_2$ grains, leading to the relatively obvious aggregation of Cu (see Figure 4c). Moreover, the release of liquid Cu during sintering caused the loss of Cu to a certain extent. Therefore, it was concluded that the optimal sintering temperature was 1000 °C for the Ti$_3$SiC$_2$/Cu composites in our study.

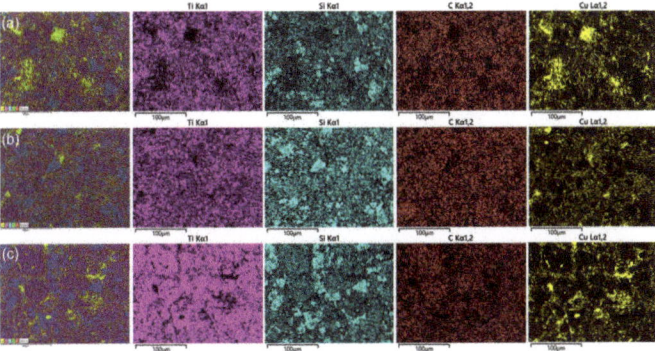

Figure 4. Mapping of elemental distribution for Ti$_3$SiC$_2$/Cu composites sintered at 950 °C (**a**), 1000 °C (**b**) and 1050 °C (**c**).

3.2. Mechanical Property

The variation in relative density, hardness and compressive strength of the Ti$_3$SiC$_2$/Cu composites with the sintered temperature are shown in Figure 5. As seen in Figure 5, the relative density of the Ti_3SiC_2/Cu composites sintered at 950 °C was 95.35%, which was slightly lower than that sintered at 1000 °C (96.43%). However, the relative density of the composite sintered at 1050 °C was 93.85%, which was the lowest among the three samples. The higher relative density of the Ti_3SiC_2/Cu composites sintered at 1000 °C was attributed to the synergetic effect of high temperature, high pressure and pulse current during the sintering process. The most important thing was that the quasi-liquid Cu possessed proper mobility, which was beneficial for the densification of the composites as well. At 1050 °C, its lowest relative density was also related to the flowability of Cu. In such circumstances, Cu flowed easily and released rapidly, causing the aggregation of Cu (see Figure 4c) and the loss of Cu (see Figure 3c).

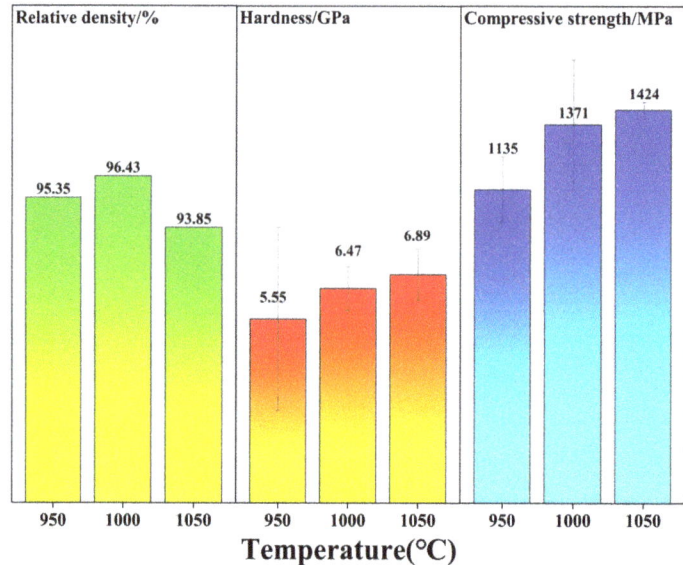

Figure 5. The variation in relative density, hardness and compressive strength of the Ti_3SiC_2/Cu composites versus the sintered temperature.

As shown in Figure 5, both the hardness and the compressive strength of the Ti_3SiC_2/Cu composites increased with the increase in the sintered temperature. It was clearly seen that the deviation of the hardness of the composites sintered at 950 °C was the highest among the three samples, which originated from the agglomeration of Cu in the composites (see Figures 3a and 4a). Compared with the polycrystalline Ti_3SiC_2 (5.5 GPa) [21], the higher hardness of the composites at different temperatures came from the formation of hard TiC product, which was detected by XRD analysis (see Figure 1).

Additionally, in comparison with the polycrystalline Ti_3SiC_2, the compressive strength of the composites sintered at different temperatures exhibited an equivalent or higher value [22], which was due to the appropriate reaction between Ti_3SiC_2 and Cu. As mentioned above, during the sintering process, the de-intercalation of Si from Ti_3SiC_2 and thereafter the dissolution of it in the liquid Cu phase led to the formation of TiC and Cu_3Si. Moreover, the Cu_3Si is uniformly distributed along the grain boundary of Ti_3SiC_2. The as-obtained fine TiC and Cu_3Si grains were uniformly distributed along the boundary of Ti_3SiC_2 grains, which was a benefit for the higher compressive strength of the composite. As seen in Figure 6a, the Ti_3SiC_2/Cu composites showed a brittle fracture character, which was identical with polycrystalline Ti_3SiC_2. From the compression fracture morphology of the composites (see Figure 6b–d), it was apparently seen that both the intergranular fracture and transgranular fracture were present on the compression fracture surface of the composites sintered at different temperatures. It indicated that the fracture mode of the composites was independent of the sintered temperature.

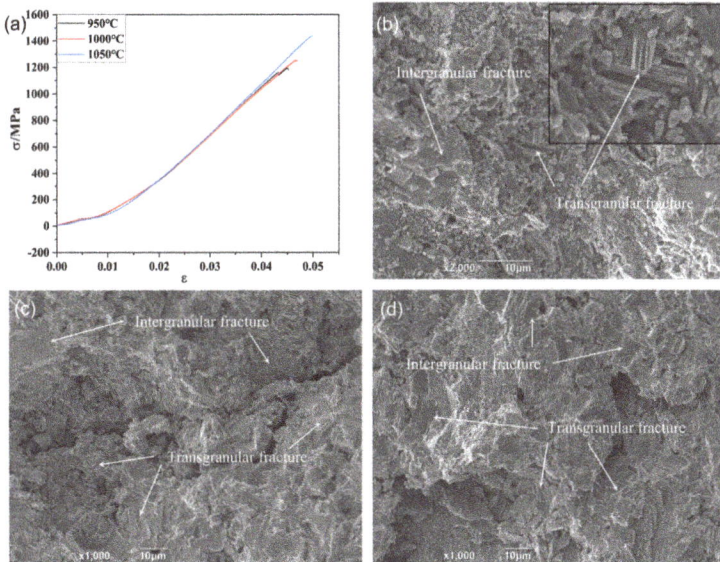

Figure 6. (**a**) Compressive stress–strain curve of Ti_3SiC_2/Cu composites, and the fracture morphology of Ti_3SiC_2/Cu composites sintered at (**b**) 950 °C (inset showed the structure at a higher magnification), (**c**) 1000 °C and (**d**) 1050 °C.

3.3. Sintering Behaviors of the Ti_3SiC_2/Cu Composites

It is instructive to explain the sintering process of the Ti_3SiC_2/Cu composites at different temperatures. The relationships of temperature, current and axial dimension with time is illustrated in Figure 7. In comparison, although the temperature difference for the three sintering temperatures was the same (950–1000 °C and 1000–1050 °C), the change in the axial dimension was different. The change in the axial dimension for 950–1000 °C was obviously lower than that for 1000–1050 °C.

The profile of temperature, current and axial dimension with time was insensitive with the sintering temperature (see Figure 7a–c). We took 1000 °C as an example to elaborate the related connections among the temperature, the current, the axial dimension and time. As seen in Figure 7b, from room temperature to 600 °C (Note: without monitoring temperature by the infrared thermometer), the heating rate was fast (approximate 100 °C/min) in order to remove moisture to dry powder completely, thus the current increased rapidly to 2300 A. During this period, the decrease in the axial dimension was due to the action of pressure and the softening due to the heating, and the first maximum of the axial dimension at about 5 min resulted from the thermal expansion and contraction. Hereafter, from 600 to the sintering temperature (for example, 1000 °C), the heating rate was set as 50 °C/min; therefore, the current rapidly adjusted to a low value (about 750 A) and then increased at a proper rate to make sure the sintering temperature intelligently controlled. At the same time, the axial dimension was controlled by a comprehensive impact of pressure, current and the thermal expansion and contraction, and it kept relatively stable before 10 min, then it remarkably decreased from 800 °C to the sintering temperature (for example, 1000 °C). The rapid decrease in the axial dimension was mainly attributed to the densification effect due to the continuous pressure at the sintering temperature and possibly due to the reaction between Ti_3SiC_2 and Cu. Finally, in the holding period, both the temperature and the current were constant, and the axial dimension slowly decreased as a result of the further densification. Additionally, an extensive reaction between Ti_3SiC_2 and Cu was partly beneficial for the decrease in the axial dimension.

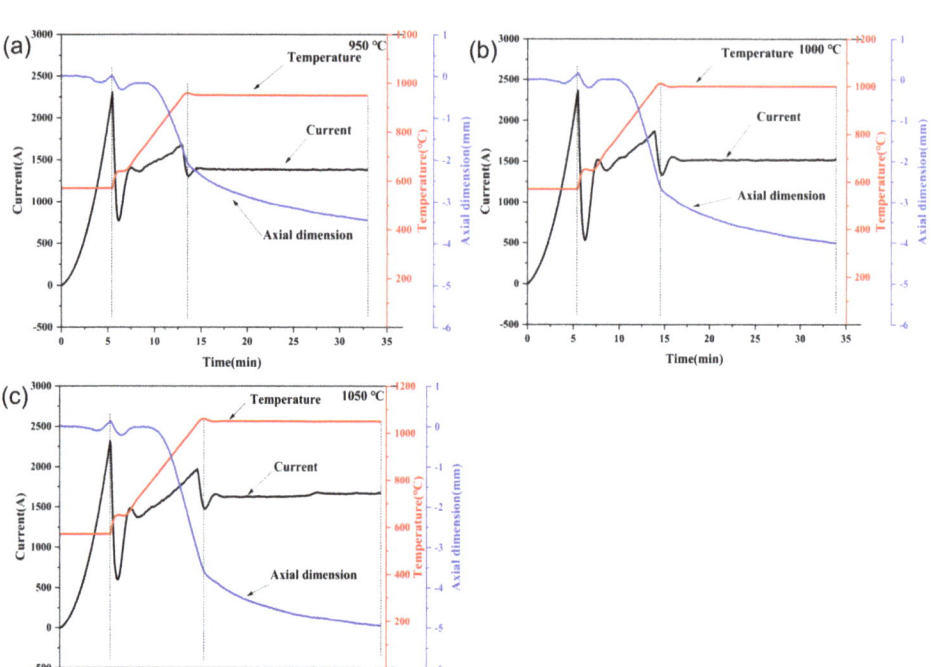

Figure 7. The relationship of the temperature, the current and the dimensional change with sintering time for the Ti_3SiC_2/Cu composites sintered at (**a**) 950 °C, (**b**) 1000 °C and (**c**) 1050 °C.

The proposed sintering process of the Ti_3SiC_2/Cu composites is shown in Figure 8. In the initial state (see Figure 8a), Ti_3SiC_2 and Cu were relatively uniformly distributed in the graphite die, which corresponded to the state of the first 10 min in Figure 7. During the initial stage, there was no reaction between Ti_3SiC_2 and Cu, and the softening of the powder and the thermal expansion and extraction were due to the action of the current and the pressure. In the transition state (see Figure 8b), under the synergetic effect of the current, temperature and pressure, Cu locally melted and was extruded into the grain boundary of Ti_3SiC_2.

On the other hand, Ti_3SiC_2 grains underwent plastic deformation under the same condition. All these factors increased the contact area of Ti_3SiC_2 and Cu. Additionally, the de-intercalation of Si atoms from Ti_3SiC_2 and its diffusion contributed to the reaction between Ti_3SiC_2 and Cu, leading to the formation of Cu_3Si and TiC, which was distributed along the grain boundary of Ti_3SiC_2 grains. Moreover, the quasi-liquid Cu diffused into the inner part of Ti_3SiC_2 grains through the holes or pores produced by the mismatch of skeleton density of the original Ti_3SiC_2 and the as-formed TiC and reacted further with the Si atoms de-intercalated from the inner Ti_3SiC_2 grains. For the Ti_3SiC_2 grain surrounded by Cu, its surface was continuously transformed into TiC, which was accompanied by the formation of Cu_3Si. Accordingly, the grain consisting of Cu_3Si and TiC, which was embedded by Ti_3SiC_2, was expected (see the inset of Figures 3b and 8b). Therefore, the grain size of Ti_3SiC_2 became smaller and smaller, which was inconsistence with the result in Figure 2. This state (the transition state in Figure 8b) corresponded to the rapid decrease in the axial dimension in Figure 7. In the final state (see Figure 8c), the reaction of Ti_3SiC_2 and Cu continued, and the pores were filled by the product, densifying the composites. This final state corresponds to the holding period in Figure 7. Consequently, the Ti$_3$SiC$_2$/Cu composites with superior mechanical properties were obtained at 1000 °C for 20 min by SPS. The proposed sintering process undoubtedly inspired us to further explore the sintering of Ti$_3$SiC$_2$/Cu composites in detail.

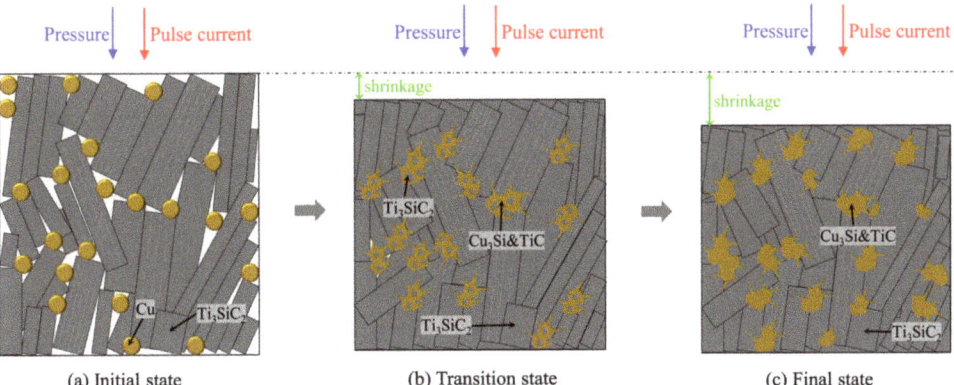

Figure 8. Proposed schematic illustration showing the sintering process of the Ti$_3$SiC$_2$/Cu composites: (**a**) initial state, (**b**) transition state and (**c**) final state.

4. Conclusions

Ti$_3$SiC$_2$/Cu composites were sintered by Spark Plasma Sintering (SPS) in a vacuum under a pressure of 35 MPa at 950–1050 °C. The as-synthesized composites were systematically characterized by scanning electron micrography (SEM), optical micrography (OM) and X-ray diffractometry (XRD), respectively. The results indicated that the constituent of the Ti$_3$SiC$_2$/Cu composites included Ti$_3$SiC$_2$, Cu$_3$Si and TiC, which was insensitive to the sintering temperature. The reaction between Ti$_3$SiC$_2$ and Cu produced Cu$_3$Si and TiC, the formation of which was attributed to the presence of Cu and the de-intercalation of Si atoms Ti$_3$SiC$_2$. OM analysis confirmed that with the increase in the sintering temperature, the Ti$_3$SiC$_2$ became smaller and smaller, which resulted from the more violent reaction between Ti$_3$SiC$_2$ and Cu. Moreover, SEM analysis demonstrated that the agglomeration of Cu in the composites limited the uniformity of the microstructure distribution of the composites, governing the mechanical behaviors of the composites. At 1000 °C, the distribution of Cu in the composites was relatively uniform; thus, the composites exhibited the highest density, relatively high hardness and compressive strength. The relationships of the temperature, the current and the axial dimension with the time during the sintering process were further discussed, and a corresponding schematic illustration was proposed to explain the related sintering behaviors of the composites sintered by SPS.

Author Contributions: Conceptualization, R.Z.; Data curation, B.C. and H.Z.; Formal analysis, R.Z., B.C., M.S. and C.W.; Investigation, R.Z., F.L., M.S. and C.W.; Methodology, R.Z. and B.C.; Writing—original draft, R.Z., B.C. and F.L.; Writing—review and editing, R.Z. and F.L. All authors have read and agreed to the published version of the manuscript.

Funding: This research was funded by the Open Project of State Key Laboratory of Solid Lubrication, Chinese Academy of Sciences (LSL-2115), and Sichuan Province Key Research and Development Program (22SYX0017).

Institutional Review Board Statement: Not applicable.

Informed Consent Statement: Not applicable.

Data Availability Statement: Not applicable.

Conflicts of Interest: The authors declare no conflict of interest.

References

1. Barsoum, M.W. The $M_{N+1}AX_N$ phases: A new class of solids: Thermodynamically stable nanolaminates. *Prog. Solid State Chem.* **2000**, *28*, 201–281. [CrossRef]
2. Zhou, Y.; Sun, Z.; Chen, S.; Zhang, Y. In-situ hot pressing/solid-liquid reaction synthesis of dense titanium silicon carbide bulk ceramics. *Mater. Res. Innov.* **1998**, *2*, 142–146. [CrossRef]
3. Barsoum, M.W.; El-Raghy, T. Synthesis and characterization of a remarkable ceramic: Ti_3SiC_2. *J. Am. Ceram. Soc.* **1996**, *79*, 1953–1956. [CrossRef]
4. El-Raghy, T.; Zavaliangos, A.; Barsoum, M.W.; Kalidindi, S.R. Damage Mechanisms around Hardness Indentations in Ti_3SiC_2. *J. Am. Ceram. Soc.* **1997**, *80*, 513–516. [CrossRef]
5. Lu, J.R.; Zhou, Y.; Zheng, Y.; Li, H.Y.; Li, S.B. Interface structure and wetting behaviour of Cu/Ti_3SiC_2 system. *Adv. Appl. Ceram.* **2015**, *114*, 39–44. [CrossRef]
6. Low, I.M. Depth profiling of phase composition in a novel Ti_3SiC_2–TiC system with graded interfaces. *Mater. Lett.* **2004**, *58*, 927–932. [CrossRef]
7. Li, H.; Peng, L.M.; Gong, M.; He, L.H.; Zhao, J.H.; Zhang, Y.F. Processing and microstructure of Ti_3SiC_2/M (M=Ni or Co) composites. *Mater. Lett.* **2005**, *59*, 2647–2649. [CrossRef]
8. Gu, W.-L.; Zhou, Y.-C. Reactions between Ti and Ti_3SiC_2 in temperature range of 1273–1573 K. *Trans. Nonferrous Met. Soc. China* **2006**, *16*, 1281–1288. [CrossRef]
9. Gu, W.L.; Yan, C.K.; Zhou, Y.C. Reactions between Al and Ti_3SiC_2 in temperature range of 600–650 °C. *Scr. Mater.* **2003**, *49*, 1075–1080. [CrossRef]
10. Kothalkar, A.D.; Benitez, R.; Hu, L.; Radovic, M.; Karaman, I. Thermo-mechanical Response and Damping Behavior of Shape Memory Alloy–MAX Phase Composites. *Metall. Mater. Trans. A* **2014**, *45*, 2646–2658. [CrossRef]
11. Dang, W.; Ren, S.; Zhou, J.; Yu, Y.; Li, Z.; Wang, L. Influence of Cu on the mechanical and tribological properties of Ti_3SiC_2. *Ceram. Int.* **2016**, *42*, 9972–9980. [CrossRef]
12. Dang, W.; Ren, S.; Zhou, J.; Yu, Y.; Wang, L. The tribological properties of Ti3SiC2/Cu/Al/SiC composite at elevated temperatures. *Tribol. Int.* **2016**, *104*, 294–302. [CrossRef]
13. Zhou, Y.C.; Gu, W.L. Chemical reaction and stability of Ti_3SiC_2 in Cu during high-temperature processing of Cu/Ti_3SiC_2 composites. *Z. Metallkd.* **2004**, *95*, 50–56. [CrossRef]
14. Zhang, R.; Liu, F.; Tulugan, K. Self-lubricating behavior caused by tribo-oxidation of Ti_3SiC_2/Cu composites in a wide temperature range. *Ceram. Int.* **2022**, *in press*.
15. Zhou, Y.; Chen, B.; Wang, X.; Yan, C. Mechanical properties of Ti_3SiC_2 particulate reinforced copper prepared by hot pressing of copper coated Ti_3SiC_2 and copper powder. *Mater. Sci. Tech. Lond.* **2004**, *20*, 661–665. [CrossRef]
16. Gao, N.F.; Miyamoto, Y. Joining of Ti_3SiC_2 with Ti–6Al–4V Alloy. *J. Mater. Res.* **2011**, *17*, 52–59. [CrossRef]
17. El-Raghy, T.; Barsoum, M.W.; Sika, M. Reaction of Al with Ti_3SiC_2 in the 800–1000 °C temperature range. *Mat. Sci. Eng. A* **2001**, *298*, 174–178. [CrossRef]
18. Tzenov, N.; Barsoum, M.W.; El-Raghy, T. Influence of small amounts of Fe and V on the synthesis and stability of Ti_3SiC_2. *J. Eur. Ceram. Soc.* **2000**, *20*, 801–806. [CrossRef]
19. Yin, X.H.; Li, M.S.; Zhou, Y.C. Microstructure and mechanical strength of diffusion-bonded Ti_3SiC_2/Ni joints. *J. Mater. Res.* **2011**, *21*, 2415–2421. [CrossRef]
20. Guo, H.; Zhang, J.; Li, F.; Liu, Y.; Yin, J.; Zhou, Y. Surface strengthening of Ti_3SiC_2 through magnetron sputtering Cu and subsequent annealing. *J. Eur. Ceram. Soc.* **2008**, *28*, 2099–2107. [CrossRef]
21. Zhang, R.; Feng, K.; Meng, J.; Liu, F.; Ren, S.; Hai, W.; Zhang, A. Tribological behavior of Ti_3SiC_2 and Ti_3SiC_2/Pb composites sliding against Ni-based alloys at elevated temperatures. *Ceram. Int.* **2016**, *42*, 7107–7117. [CrossRef]
22. Zhang, R.; Feng, K.; Meng, J.; Su, B.; Ren, S.; Hai, W. Synthesis and characterization of spark plasma sintered Ti_3SiC_2/Pb composites. *Ceram. Int.* **2015**, *41 Pt A*, 10380–10386. [CrossRef]

Article

Microstructure and Tribological Properties of Spark-Plasma-Sintered Ti$_3$SiC$_2$-Pb-Ag Composites at Elevated Temperatures

Rui Zhang [1,2,3,4,*], Huiming Zhang [1] and Fuyan Liu [5,*]

1. School of Mechanical Engineering, Chengdu University, Chengdu 610106, China; zhm121998@163.com
2. Sichuan Province Engineering Technology Research Center of Powder Metallurgy, Chengdu University, Chengdu 610106, China
3. Institute for Advanced Materials Deformation and Damage from Multi-Scale, Chengdu University, Chengdu 610106, China
4. School of Mechanical Engineering, Xinjiang University, Urumqi 830000, China
5. School of Chemical Engineering and Materials, Changzhou Institute of Technology, Changzhou 213032, China
* Correspondence: zhangrui0214@cdu.edu.cn (R.Z.); liufy@czu.cn (F.L.)

Abstract: Ti$_3$SiC$_2$-PbO-Ag composites (TSC-PA) were successfully prepared using the spark plasma sintering (SPS) technique. The ingredient and morphology of the as-synthesized composites were elaborately investigated. The tribological properties of the TSC-PA pin sliding against Inconel 718 alloys disk at room temperature (RT) to 800 °C were examined in air. The wear mechanisms were argued elaborately. The results showed that the TSC-PA was mainly composed of Ti$_3$SiC$_2$, Pb, and Ag. The average friction coefficient of TSC-PA gradually decreased from 0.72 (RT) to 0.3 (800 °C), with the temperature increasing from RT to 800 °C. The wear rate of TSC-PA showed a decreasing trend, with the temperature rising from RT to 800 °C. The wear rate of Inconel 718 exhibited positive wear at RT and negative wear at elevated temperatures. The tribological property of TSC-PA was related to the tribo-chemistry, and the abrasive and adhesive wear.

Keywords: Ti$_3$SiC$_2$-PbO-Ag composites; tribo-chemical reaction; self-lubricating composite; tribo-oxidation

1. Introduction

Ti$_3$SiC$_2$ belongs to one of the MAX phases (polycrystalline nanolaminates of ternary carbides and nitrides, over 60 + phases) [1]. It possesses the combined characteristics of metal and ceramics, such as being easily machinable, being electrically and thermally conductive, being oxidation resistant, and having a high melting point. Ti$_3$SiC$_2$ can be a potential candidate for structural functional components. Similar to graphite and MoS$_2$, it has a hexagonal structure, endowing it with an exceptional lubricating property. Research on the tribological properties of Ti$_3$SiC$_2$ demonstrated that it is not a special solid lubricant. Contrarily, Ti$_3$SiC$_2$ generally has a high friction coefficient ($\mu > 0.6$) and a high wear rate (WR: in the order of 10^{-2}–10^{-3} mm^3/N m), especially at medium and low temperatures (RT) [2–8]. It has been reported that, in some circumstances (such as with temperature > 600 °C, at a speed of >5 m/s, or sliding against some particular tribopair material), polycrystalline Ti$_3$SiC$_2$ displays the relatively low μ and wear rate, which mainly results from the tribo-oxidation transfer films generated on the contact surface [2,6,8–10]. In other situations, the fracture and pulling-out of Ti$_3$SiC$_2$ grains causes three-body abrasive wear, which is the main wear mechanisms for Ti$_3$SiC$_2$.

Subsequently, most of scientific enthusiasts concentrate on the intensifying research of polycrystalline Ti$_3$SiC$_2$ by uniting metal and/or ceramic to enhance its friction and wear behaviors. Ren et al. [11] prepared Ti$_3$SiC$_2$/Cu/Al/SiC composites, and they found that the tribological behaviors of the composite were better than those of polycrystalline Ti$_3$SiC$_2$ at RT and 200 °C, while the wear properties of the composite were worse than that of polycrystalline Ti_3SiC_2 at elevated temperatures (≥ 400 °C). The incorporation of Al, Cu, SiC and Al_2O_3 played an important role in reinforcing the bonding of Ti_3SiC_2 grains and in fixing the soft Ti_3SiC_2 matrix around them during reciprocate sliding at RT–200 °C, where the wear mechanism was abrasive wear. At elevated temperatures (≥ 400 °C), the plastic flow of tribo-oxides layers led to the higher wear rate of the composite than that of Ti_3SiC_2 where the wear mechanism was adhesive wear. Yang et al. [12] prepared Cu-Ti_3SiC_2 co-continuous composites. They showed excellent electrical conductivity and wear resistance due to the addition of Cu. Yang et al. [13] explored the tribological behaviors of (TiB_2 + TiC)/Ti_3SiC_2 composites. It was discovered that μ of the composites was bigger compared with that of Ti_3SiC_2. The addition of TiB_2 and TiC could fix the soft matrix around them and scatter the shear stresses. In our earlier work [14], we discussed the tribological behavior of Ti_3SiC_2/CaF_2 composites at elevated temperatures. The prepared Ti_3SiC_2/CaF_2 exhibited better friction and wear property than Ti_3SiC_2 in a wide temperature range due to the tribo-oxide competition. Islak et al. [15] investigated the effect of reinforcing TiB_2 particles (5, 10, and 15 wt.%) on the properties of as-synthesized Ti_3SiC_2. The addition of TiB_2 improved the mechanical properties and thermal diffusivity of the composites. Moreover, the addition of TiB2 significantly increased the wear properties of Ti_3SiC_2 matrix. Magnus et al. [16] prepared the Ti_3SiC_2-$TiSi_2$-TiC composites by spark plasma sintering (SPS). They discovered that the tribological property of the composites was attributed to the intrinsic lubricity and the addition of second phase TiC particles. During the sliding, the tribo-oxidative wear was the main wear mechanism; then, it changed to the deformation-induced wear. The transition resulted from the breakdown of the formed tribofilm.

Silver shows some outstanding characteristics (such as good thermal conductivity, favorable ductility, and special mechanical behaviors) and can be widely applied in many fields, such as air-foil bearings, and biomedical and thermal interface materials. It is commonly utilized as a solid lubricant at temperatures below 500 °C. Additionally, it has a high diffusion coefficient and easily generates a lower shear stress intersection at the sliding interface [17,18]. Thus, it was significant and practical to incorporate good properties of Ag with Ti_3SiC_2. The goal of this work was to synthesize Ti_3SiC_2/Ag composites and to uncover the mechanical and tribological behaviors of the composite at raising temperatures. F. AlAnazi et al. [19] prepared metals-based (Ag and Bi) composites incorporation with 5 vol%, 10 vol%, 20 vol%, and 30 vol% Ti_3SiC_2 and investigated their tribological properties. They found that the incorporation of Ti_3SiC_2 improved the hardness and compressive yield strength of the two composites. Additionally, the addition of Ti_3SiC_2 enhanced the tribological performance of the two composites. Zeng et al. [20] inspected the tribological behaviors of Ti_3SiC_2/Ag composites with Inconel 718 as the tribopair at RT. At different sliding speeds, the composites showed lower μ and wear rate than that of Ti_3SiC_2.

Pb is lubricous due to its low shear strength. In our previous work [21,22], we found that the incorporation of PbO (melting point: 888 °C) instead of Pb (melting point: 327 °C) in Ti_3SiC_2 can successfully prepare Ti_3SiC_2/Pb composites and avoid the loss of Pb during the sintering process. The as-prepared composite showed better friction and wear behaviors than Ti_3SiC_2 at elevated temperatures.

As a result of the above-mentioned facts, it was necessary to suggest some novel composites that can move forward a single step in strengthening the tribological performance of Ti_3SiC_2 in a wide temperature range. Therefore, in this work, the metal Pb was added to the Ti_3SiC_2/Ag composite to prepare Ti_3SiC_2-Pb-Ag composites (TSC-PA). To prevent the loss of Pb during the process of preparing the TSC-PA, PbO was chosen as the source of Pb. The target of this job was to synthesize TSC-PA and to discuss the mechanical and tribological behaviors of the as-synthesized compound at elevated temperatures.

2. Experimental Procedure
2.1. Materials

The TSC-PA composites were sintered using the powder mixture of Ti_3SiC_2, PbO and Ag. The average particle size of Ti_3SiC_2 was 3 μm and its purity was $\geq 98\%$. It was purchased from Jinhezhi Materials Ltd., Beijing, China. The average particle size of PbO and Ag was 5 μm, and their purity was ≥99%. They were bought from Xilong Chemical Ltd., Shantou, China. The powder composition was 70 vol% Ti_3SiC_2-15 vol%PbO-15 vol%Ag. The mixture with a designated composition was blended by ball mill for 6 h. Next, it was loaded in a graphite die. In the end, it was sintered by SPS furnace (Shanghai Chenhua Electric Furnace Co., Ltd., Shanghai, China) in vacuum at 1130 °C.

2.2. Mechanical Properties

The densities of the composites were determined using Archimedes's principle. Microhardness was measured on a Micro-hardness Tester (MH-5-VM, Shanghai Hengyi Technology Co., Ltd., Shanghai, China). with a load of 500 g and a dwell time of 10 s. Flexural strength and compression strength were measured using a universal material tester (SANS-CMT5205, Shenzhen New Sansi Material Testing Co., Ltd., Shenzhen, China). The three-point bending test was conducted to obtain the flexural strength, and the samples were cut to 3 mm high × 4 mm wide × 20 mm long. The cross-head speed and span were 0.05 mm/min and 16 mm, respectively. The sizes of the samples for the compression strength test were Φ 5 mm × 12.5 mm, and the cross-head speed was 0.2 mm/min.

2.3. Friction and Wear Test

The friction and wear experiments were carried out in air on a high-temperature tribometer with a pin-on-disk setup (THT01-04015, CSM Instruments SA, Peseux, Switzerland). The TSC-PA was used as a pin, the size of which was Φ 6 mm × 12 mm. The Inconel 718 was used as a disk, and its size was Φ 32 mm × 8 mm. The surfaces of the disk and pin were ground and polished into a surface roughness (Ra) of 0.06 μm. The tribopairs slid against each other for 200 m at a sliding speed of 0.1 m/s under a normal load of 5 N. The friction coefficient was automatically documented by computer during the experiment. The wear volume was obtained by quantifying the volume loss of TSC-PA pin by optical microscopy and by examining the cross-sectional area of the worn area of Inconel 718 disk by 3D surface profilometry (NanoMap-D, Columbus, OH, USA). The wear rates were acquired by Equation (1).

$$\text{Wear rate} \left(\frac{\text{mm}^3}{\text{N·m}}\right) = \frac{\text{Wear volume } (\text{mm}^3)}{\text{Sliding distance (m)} \times \text{Normal load (N)}} \quad (1)$$

2.4. Analysis

The morphology of TSC-PA was investigated by scanning electron microscopy (SEM, JSM-5600LV, JEOL, Tokyo, Japan) equipped with energy dispersive X-ray spectroscopy. The EDS (IE250, Oxford Instrument, Abingdon, Oxfordshire, UK) was equipped with X-Max Silicon Drift Detector (SDD; the spectral resolution of 124 eV at Mn Kα) and an ultra-thin window; the active area of the SDD was 50 mm². The phase composition of TSC-PA was analyzed by X-ray diffraction (XRD, Philips X'Pert Pro, PANalytical Netherlands, Almelo, Netherland) using Cu Kα radiation (λ = 0.15418 nm) within the 2θ-angle range from 5° to 90°. X-ray photoelectron spectroscope (XPS, PHI-5702, Physical Electronics Corporation, Chanhassen, MN, USA) was used to analyze the elemental chemical states of TSC-PA on the worn surfaces, and the binding energy of adventitious carbon (C1s: 284.8 eV) was used as the reference.

3. Result and Discussion

3.1. Phase Composition and Microstructure of TSC-PA

The XRD pattern of TSC-PA is displayed in Figure 1. As noticed in Figure 1, the main phases of TSC-PA were Ti_3SiC_2, Pb, and Ag. The Ti_3SiC_2 was temporarily stable and did not break down during the sintering process of TSC-PA. To the best of the authors' knowledge, the Ti_3SiC_2 revealed high reaction activity when in contact with the metal phases, such as Cu [23–25], Al [26,27], and Fe [28]. In this study, the Ti_3SiC_2 did not decompose with the co-addition of Ag and PbO at the sintering temperature of 1130 °C. In a previous paper [21], we reported that PbO in the Ti_3SiC_2-PbO system was deoxidized to Pb by C or Si during the preparation process and Ti_3SiC_2 did not decompose at 1200 °C. The relevant chemical reactions were described in Reference [21]. In comparison, the addition of Ag did not change the composition of the Ti_3SiC_2-PbO system.

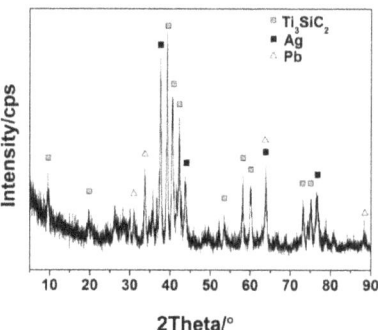

Figure 1. The XRD pattern of TSC-PA.

As observed in Figure 2, the morphology and the distributions of elements of the cross section of the TSC-PA was analyzed using SEM. It can be observed that the Pb particles were homogeneously scattered in the Ti_3SiC_2 matrix, while the Ag particles were agglomerated to a certain extent. As the melting point of silver was merely 961 °C, some amount of molten silver nearby flowed together during sintering. Therefore, the agglomeration of Ag occurred in the TSC-PA.

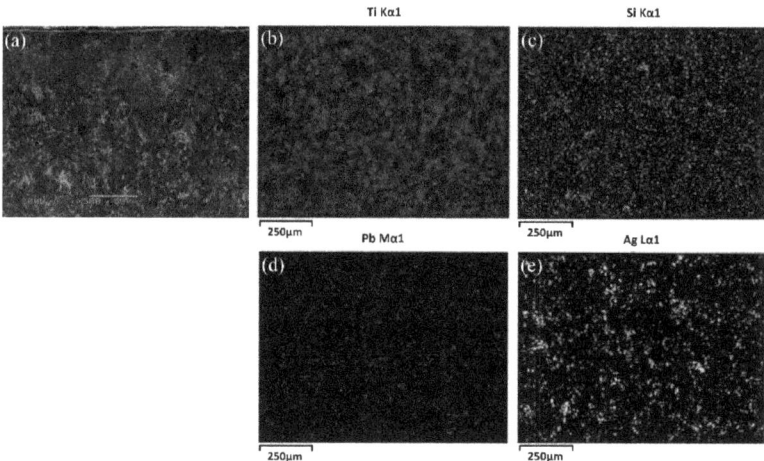

Figure 2. SEM micrograph (**a**) and distribution of Ti (**b**), Si (**c**), Pb (**d**), Ag (**e**) elements in the cross section of TSC-PA.

3.2. Mechanical Performance of TSC-PA

The relative density, microhardness, flexural strength, and compressive strength of TSC-PA and TSC are listed in Table 1. It can be seen that the relative densities of TSC-PA and TSC were 95.35% and 98.24%, respectively. The microhardness of TSC-PA was less than half that of TSC, and their microhardness values were 2.24 and 5.5 GPa, respectively. Pb and Ag in the TSC-PA belong to soft metals and caused the low hardness of TSC-PA. The flexural strength of TSC-PA was less than that of TSC, and their values were 311 and

428 MPa, respectively. The compressive strength of TSC-PA was only half that of TSC, and their values were 654 MPa and 1230 MPa, respectively.

Table 1. Properties of TSC and TSC-PA.

Sample	No.	Relative Density /%	Microhardness /GPa	Flexural Strength /MPa	Compressive Strength /MPa
Ti_3SiC_2 (1250 °C)	TSC	98.24	5.5 ± 0.2	428 ± 10	1230 ± 13
Ti_3SiC_2-PbO-Ag	TSC-PA	95.35	2.24 ± 0.14	311 ± 22	654

Note: The data in the first line were from Reference [21].

As shown in Figure 3a, the TSC-PA only suffered from elastic deformation before fracture, which was a typical brittle fracture. The fracture surface of TSC-PA (see Figure 3b) indicated that the fracture surface of TSC-PA was featured by intergranular and transgranular fracture. This fracture mode of TSC-PA was similar to TSC. As well known, the diverse energy-absorbing strategies including the diffuse of the microcracks, lamination, crack deflection, and grain pull-out were discovered to be in charge of the high flexural strength of TSC [29,30]. In addition to intergranular and transgranular fracture, some pores and cracks resulting from the unsatisfactory compatibility (e.g., wettability) of the TSC matrix and Pb and Ag were observed (see Figure 3b). The thermal expansion coefficients (for instance: TSC: 9.1×10^{-6}, Pb: 29.3×10^{-6}, and Ag: 19.5×10^{-6} °C^{-1}) of different phases were not matched with each other, leading to the weak bonding between granules and the formation of some defects (such as micro-pores and micro-cracks) during the preparation procedure. The as-formed defects were detrimental to the flexural strength. An additional major justification for the poor fracture strength of the TSC-PA was that the distribution of Pb and Ag around the TSC grains to a certain extent hindered the diverse energy-assimilating strategies, which let fracture take place between the grains.

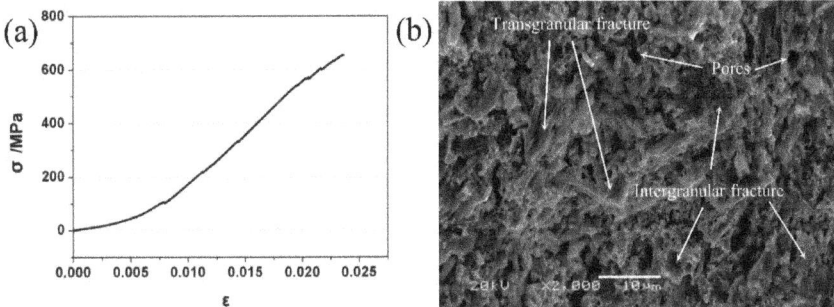

Figure 3. (a) Compressive stress–strain curve of TSC-PA and (b) SEM image of the fractured surface of TSC-PA after the three-point bending test.

3.3. Tribological Behavior of TSC-PA

Figure 4a showed the plot of μ versus distance curve of the TSC-PA/Inconel 718 tribopair at elevated temperatures. It can be easily construed that the temperature had a direct effect on the μ of the tribopair. At RT and 200 °C, the μ had a big fluctuation. At 400, 600, and 800 °C, the incipient μ had a big fluctuation and, then, it soon achieved a steady state with low μ values. Specifically, at RT, the μ showed a larger fluctuation than that at other temperatures. Particularly, at 600 °C, the μ exhibited a turbulent behavior in contrasted with that at other temperatures, which indicated an incipient short break-in period with relatively low μ values; then, it reached high μ values and fluctuated largely. Later, it approached a stable state with relatively low μ values. In comparison, the fluctuation of the μ decreased with the temperature increasing from RT to 800 °C. Figure 4b compared the μ_{mean} of TSC-PA/Inconel 718 tribopair at elevated temperatures. It can be seen that the μ_{mean} decreased with the temperature increasing from RT to 800 °C.

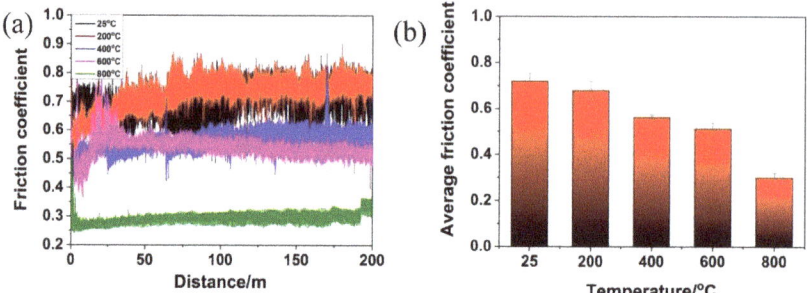

Figure 4. Plot of (**a**) friction coefficient (μ) versus distance and (**b**) the average friction coefficients of TSC-PA sliding against Inconel 718 alloys with different temperature.

Figure 5a plotted the wear rates of the TSC-PA pin at elevated temperatures. The wear rates of the pin decreased with the temperature increasing from RT to 400 °C. Moreover, the wear rates of the pin were the same when the temperature was at 400, 600, and 800 °C. No other results were found in the literature for a comparison with the results in this work because Ti_3SiC_2 composites with the co-addition of PbO and Ag have not been reported before. Moreover, it was not easy to make a comparison between tribology results as they were carried out under distinct conditions on the basis of the detailed demands of an application. For interpretative aspirations, Table 2 compared tribological behavior of Ti_3SiC_2 composites with the addition of the different amount of Ag or the addition of the different amount of PbO. Comparatively, the incorporation of relative low content (5–20 vol%) of Ti_3SiC_2 can diminish the wear rate of the Ag-based composite during sliding against Al_2O_3 at 5N and 0.5 m/s [19]. In this work, the situation was different because the sample was Ti_3SiC_2 based-composite with 70 vol% of Ti_3SiC_2. At RT, comparatively, Ti_3SiC_2 and composites of Ti_3SiC_2 with 15 vol% Ag, with 15 vol% PbO, and with 15 vol% PbO-15 vol% Ag showed wear rates of 2×10^{-3} mm^3/(N·m), 3×10^{-5} mm^3/(N·m), 2×10^{-4} mm^3/(N·m), and 2×10^{-4} mm^3/(N·m), respectively, during sliding against Inconel 718 at 5N and 0.1 m/s [20,22]. These results showed that the incorporation of 15 vol% Ag or 15 vol% PbO or 15 vol% PbO-15 vol% Ag can decrease the wear rate of Ti_3SiC_2 by at least an order of magnitude at RT. At elevated temperatures (>200 °C), the composite of Ti_3SiC_2 with 15 vol% PbO showed better wear resistance than both Ti_3SiC_2 and the composite of Ti_3SiC_2 with 15 vol% PbO-15 vol% Ag (TSC-PA). The tribological behavior of the composite of Ti_3SiC_2 with 15 vol% Ag at elevated temperature was unavailable and incomparable. It can be seen from Table 2 that TSC-PA showed a lower friction than TSC at > 200 °C and a lower friction than Ti_3SiC_2-15 vol% PbO at <600 °C. It was concluded that the addition of PbO and Ag apparently improved the tribological behavior of TSC-PA, in comparison with TSC and Ti_3SiC_2-15 vol% PbO.

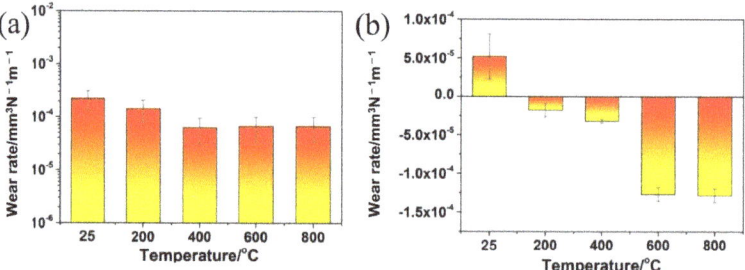

Figure 5. The wear rates of (**a**) pin and (**b**) disk for TSC—PA/Inconel 718 alloy tribo-pair as a function of temperature.

Table 2. Comparison of tribological behavior of different Ti_3SiC_2-based composites.

Composition	Counter-Surface	Conditions	Wear rate (mm^3/Nm)	μ	Reference
Ag-5 vol % Ti_3SiC_2		Block (Tab)-on-Disc, 5N, 0.5 m/s, air, RT	3.2×10^{-5}	0.38	
Ag-10 vol % Ti_3SiC_2		Block (Tab)-on-Disc, 5N, 0.5 m/s, air, RT	2.9×10^{-5}	0.35	
Ag-20 vol % Ti_3SiC_2	Al_2O_3	Block (Tab)-on-Disc, 5N, 0.5 m/s, air, RT	4.1×10^{-6}	0.29	[19]
Ag-30 vol % Ti_3SiC_2		Block (Tab)-on-Disc, 5N, 0.5 m/s, air, RT	2.5×10^{-5}	0.3	
Ag		Block (Tab)-on-Disc, 5N, 0.5 m/s, air, RT	4.9×10^{-5}	0.38	
Ti_3SiC_2		Pin-on-disk, 5N, 0.01 m/s, air, RT	4.0×10^{-3}	0.62	
		Pin-on-disk, 5N, 0.1 m/s, air, RT	2.0×10^{-3}	0.61	
		Pin-on-disk, 5N, 1 m/s, air, RT	6.0×10^{-3}	0.57	
Ti_3SiC_2-5 vol% Ag		Pin-on-disk, 5N, 0.01 m/s, air, RT	7.0×10^{-5}	0.42	
		Pin-on-disk, 5N, 0.1 m/s, air, RT	8.0×10^{-5}	0.51	
		Pin-on-disk, 5N, 1 m/s, air, RT	5.0×10^{-3}	0.56	
Ti_3SiC_2-10 vol% Ag	Inconel 718	Pin-on-disk, 5N, 0.01 m/s, air, RT	4.0×10^{-5}	0.4	[20]
		Pin-on-disk, 5N, 0.1 m/s, air, RT	7.0×10^{-6}	0.51	
		Pin-on-disk, 5N, 1 m/s, air, RT	3.0×10^{-3}	0.5	
Ti_3SiC_2-15 vol% Ag		Pin-on-disk, 5N, 0.01 m/s, air, RT	5.0×10^{-5}	0.4	
		Pin-on-disk, 5N, 0.1 m/s, air, RT	3.0×10^{-5}	0.5	
		Pin-on-disk, 5N, 1 m/s, air, RT	2.0×10^{-5}	0.45	
Ti_3SiC_2-20 vol% Ag		Pin-on-disk, 5N, 0.01 m/s, air, RT	2.0×10^{-5}	0.31	
		Pin-on-disk, 5N, 0.1 m/s, air, RT	2.0×10^{-4}	0.45	
		Pin-on-disk, 5N, 1 m/s, air, RT	1.0×10^{-5}	0.4	
Ti_3SiC_2	Inconel 718 disk	Pin-on-disk, 5N, 0.1 m/s, air, RT	2.0×10^{-3}	0.65	
		Pin-on-disk, 5N, 0.1 m/s, air, 200 °C	3.0×10^{-4}	0.65	
		Pin-on-disk, 5N, 0.1 m/s, air, 400 °C	5.0×10^{-5}	0.62	
		Pin-on-disk, 5N, 0.1 m/s, air, 600 °C	2.5×10^{-5}	0.65	
		Pin-on-disk, 5N, 0.1 m/s, air, 800 °C	5.0×10^{-6}	0.4	
Ti_3SiC_2-5 vol% PbO	Inconel 718 disk	Pin-on-disk, 5N, 0.1 m/s, air, RT	2.0×10^{-3}	0.72	
		Pin-on-disk, 5N, 0.1 m/s, air, 200 °C	1.5×10^{-3}	0.66	
		Pin-on-disk, 5N, 0.1 m/s, air, 400 °C	9.0×10^{-5}	0.65	
		Pin-on-disk, 5N, 0.1 m/s, air, 600 °C	4.0×10^{-5}	0.58	[22]
		Pin-on-disk, 5N, 0.1 m/s, air, 800 °C	7.0×10^{-6}	0.45	
Ti_3SiC_2-10 vol% PbO	Inconel 718 disk	Pin-on-disk, 5N, 0.1 m/s, air, RT	6.0×10^{-4}	0.67	
		Pin-on-disk, 5N, 0.1 m/s, air, 200 °C	6.0×10^{-4}	0.65	
		Pin-on-disk, 5N, 0.1 m/s, air, 400 °C	3.0×10^{-5}	0.66	
		Pin-on-disk, 5N, 0.1 m/s, air, 600 °C	1.0×10^{-5}	0.46	
		Pin-on-disk, 5N, 0.1 m/s, air, 800 °C	5.0×10^{-6}	0.3	
Ti_3SiC_2-15 vol% PbO	Inconel 718 disk	Pin-on-disk, 5N, 0.1 m/s, air, RT	2.0×10^{-4}	0.71	
		Pin-on-disk, 5N, 0.1 m/s, air, 200 °C	2.0×10^{-4}	0.75	
		Pin-on-disk, 5N, 0.1 m/s, air, 400 °C	1.0×10^{-5}	0.65	
		Pin-on-disk, 5N, 0.1 m/s, air, 600 °C	5.0×10^{-6}	0.35	
		Pin-on-disk, 5N, 0.1 m/s, air, 800 °C	2.0×10^{-6}	0.2	
Ti_3SiC_2-15 vol%PbO-15 vol % Ag	Inconel 718 disk	Pin-on-disk, 5N, 0.1 m/s, air, RT	2.0×10^{-4}	0.72	
		Pin-on-disk, 5N, 0.1 m/s, air, 200 °C	1.5×10^{-4}	0.67	
		Pin-on-disk, 5N, 0.1 m/s, air, 400 °C	6.0×10^{-5}	0.56	This work
		Pin-on-disk, 5N, 0.1 m/s, air, 600 °C	6.0×10^{-5}	0.51	
		Pin-on-disk, 5N, 0.1 m/s, air, 800 °C	6.0×10^{-5}	0.3	

Figure 5b compared the wear rates of the Inconel 718 disk sliding against TSC-PA at elevated temperatures. As the temperature raised from RT to 800 °C, the wear rates of the disk decreased. At RT, a positive wear rate was found for the disk. Interestingly, negative wear rates were detected for the disk and increased with the temperature rising from 200 to 800 °C. Moreover, the wear rate of the disk at 600 °C was identical to that at 800 °C. The detection of Ti, Si, Pb, and Ag on the worn surface of Inconel 718 was in line with the negative wear of the disk (see Table 3).

3.4. Tribological Mechanisms

Figure 6 showed the SEM assessment of tribosurfaces of TSC-PA/Inconel 718 tribocouple after tribological testing at elevated temperature. TSC-PA was full of scars resulting from the abrasive wear, while the Inconel 718 surface was covered with an even tribolayer delivered from the TSC-PA surface. The surface was surrounded by the tribolayer made up of incompletely oxidized TSC-PA (Table 3 and Figure 6). With the rise in temperature, the tribofilm formed on the surface of TSC-PA was smoother and denser. This film was formed undoubtably by a coalescence of diverse mechanical blending, fracturing, resintering, and mechanically vitalized oxidation and other chemical processes, perhaps, continuously refined during the heat cycle. The EDS data showed that the elements of Ni, Cr, and Fe from Inconel 718 alloy appeared on the surface of TSC-PA (Figure 6a,c,e,g,i in Table 3) and the elements of Ti, Si, Pb, and Ag from TSC-PA appeared on the surface of Inconel718 alloy (Figure 6b,d,f,h,j in Table 3). Moreover, tribo-oxidation was a predominant factor that influenced, and even controlled, the wear mechanism and finally brought about the change of friction and wear features [31]. The chemical states of Ti, Si, Pb, and Ag elements on the worn surface of TSC-PA at RT–800 °C were investigated in detail by XPS (see Figure 7).

Table 3. The chemical components of samples corresponding to Figure 6 as determined by EDS.

Positions	Samples	Temperature/°C	Atomic Percentage/at. %
Figure 5a	TSC-PA pin	25	14.9%Ti, 5.2%Si, 29.9%C, 1.4%Al, 1.9%Ag, 42.8%O, 1.0%Pb, 1.3%Ni, 0.7%Cr, 0.9%Fe
Figure 5b	Inconel 718 disk	25	2.4%Ti, 1.3%Si, 28.8%C, 0.9%Al, 0.4%Ag, 25.4%O, 21.0%Ni, 9.4%Cr, 8.4%Fe, 1.3%Nb, 0.7%S
Figure 6c	TSC-PA pin	200	17.3%Ti, 6.3%Si, 31.2%C, 1.3%Al, 1.6%Ag, 39.5%O, 1.0%Pb, 0.8%Ni, 0.5%Cr, 0.5%Fe
Figure 6d	Inconel 718 disk	200	6.5%Ti, 3.3%Si, 31.7%C, 1.4%Al, 0.9%Ag, 33.6%O, 0.5%Pb, 11.1%Ni, 5.3%Cr, 4.6%Fe, 0.7%Nb, 0.4%S
Figure 6e	TSC-PA pin	400	20.5%Ti, 6.1%Si, 23.4%C, 2.0%Al, 2.0%Ag, 42.0%O, 1.2%Pb, 1.5%Ni, 0.6%Cr, 0.7%Fe
Figure 6f	Inconel 718 disk	400	13.0%Ti, 5.7%Si, 18.2%C, 1.9%Al, 2.0%Ag, 46.4%O, 1.0%Pb, 6.1%Ni, 3.0%Cr, 2.7%Fe
Figure 6g	TSC-PA pin	600	9.5%Ti, 3.0%Si, 11.0%C, 0.8%Al, 0.5%Ag, 61.0%O, 0.4%Pb, 7.4%Ni, 3.2%Cr, 2.7%Fe, 0.5%Nb
Figure 6h	Inconel 718 disk	600	2.3%Ti, 0.7%Si, 15.4%C, 0.5%Al, 45.3%O, 1.1%Pb, 17.9%Ni, 7.8%Cr, 6.9%Fe, 1.3%Nb, 0.8%S
Figure 6i	TSC-PA pin	800	9.7%Ti, 2.9%Si, 10.7%C, 0.9%Al, 0.7%Ag, 59.7%O, 0.6%Pb, 7.8%Ni, 3.5%Cr, 2.9%Fe, 0.6%Nb
Figure 6j	Inconel 718 disk	800	11.9%Ti, 5.9%Si, 18.7%C, 1.7%Al, 2.3%Ag, 46.9%O, 0.9%Pb, 5.9%Ni, 3.2%Cr, 2.6%Fe

At RT, some wear debris dispersed on the worn surface of the TSC-PA with several dark continent regions (see Figure 6a). It was worth mentioning that some loose cracks and pores scattered on the worn surface, which was due to the peeling-off and decentralization of the granules. Such an episode was in accordance with Ti_3SiC_2 [6]. Therefore, it can be deduced that the addition of Pb and Ag did not affect the wear behavior of Ti_3SiC_2 at RT. Though the weak bonding between Ti_3SiC_2 grains in the TSC-PA led to the fracture and pulling out of Ti_3SiC_2 granules, soft metals (including Pb and Ag) distributed around the Ti_3SiC_2 grains, inhibiting the peeling-off and decentralization of Ti_3SiC_2 matrix; therefore, the wear rate of TSC-PA was only one tenth of Ti_3SiC_2 (see Table 2). A third body was formed and entrapped between the pin and disk, leading to the severe wear of TSC-PA and Inconel 718 at RT. The obvious evidence was that wear grooves were found on the worn surface of Inconel 718 after tribology testing at RT (see Figure 6b) and mutual material transfer between the pin and disk was detected by EDS analysis (see Table 3). The abrasive wear was the primary wear mechanism for TSC-PA. A small amount of SiO_2 and PbO was detected on the worn surface of TSC-PA (see Figure 7). However, the temperature was too low to form an enough thick tribo-oxidation film between the contact surface of TSC-PA and Inconel 718, so high friction coefficients and high wear rates were present at RT (see Figures 4 and 5a).

At 200 °C, some compacted wear debris was found on the worn surface of TSC-PA (see Figure 6c). Compared with RT, besides SiO_2 and PbO, other oxides, TiO_2 and Ag_2O, were formed on the surface of TSC-PA (evidently owing to the XPS at 200 °C in Figure 7), generating a thicker tribo-oxidation film. The formation of the tribo-oxidation film did not inhibit the fracture and pulling-out of Ti_3SiC_2 grains. The formation and entrapment of the third body (the formed wear debris) between the pin and disk contributed to the wear of TSC-PA and Inconel 718. The mutual material transfer between the pin and disk at 200 °C was detected by EDS analysis (see Table 3). A tribo-oxidation film with some wear grooves was transferred to the Inconel 718, so the counterpart showed relatively flat region and slight negative wear (see Figures 5b and 6d). At 200 °C, the abrasive wear mechanism was the main wear mechanism with very slight adhesive wear.

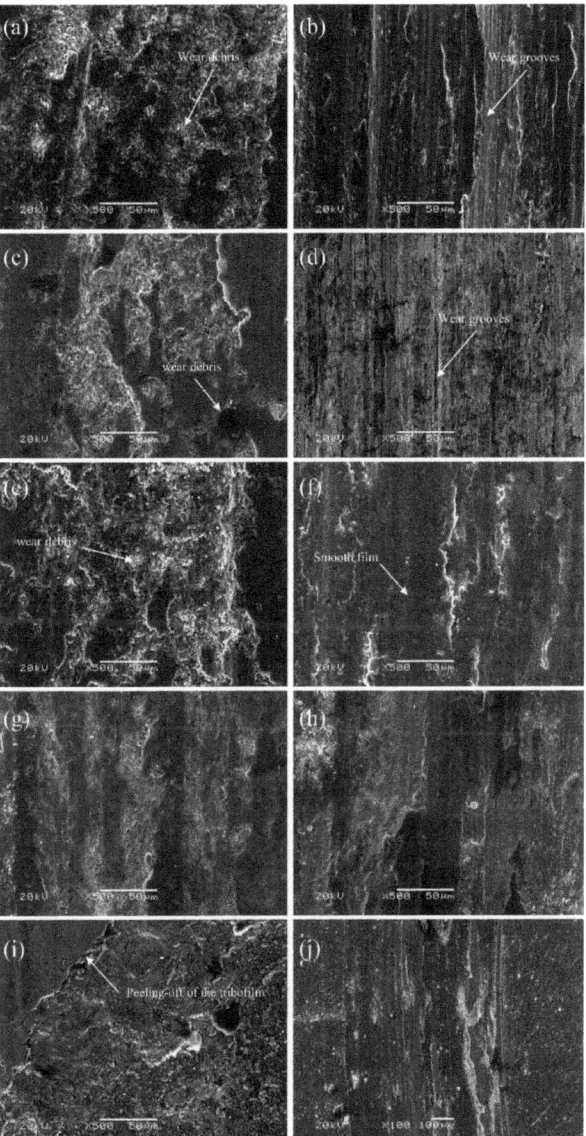

Figure 6. SEM micrographs of (**a**) TSC-PA at 25 °C, (**b**) Inconel 718 alloy at 25 °C, (**c**) TSC-PA at 200 °C, (**d**) Inconel 718 alloy at 200 °C, (**e**) TSC-PA at 400 °C, (**f**) Inconel 718 alloy at 400 °C, (**g**) TSC-PA at 600 °C, (**h**) Inconel 718 alloy at 600 °C, (**i**) TSC-PA at 800 °C, and (**j**) Inconel 718 alloy at 800 °C after tribological testing.

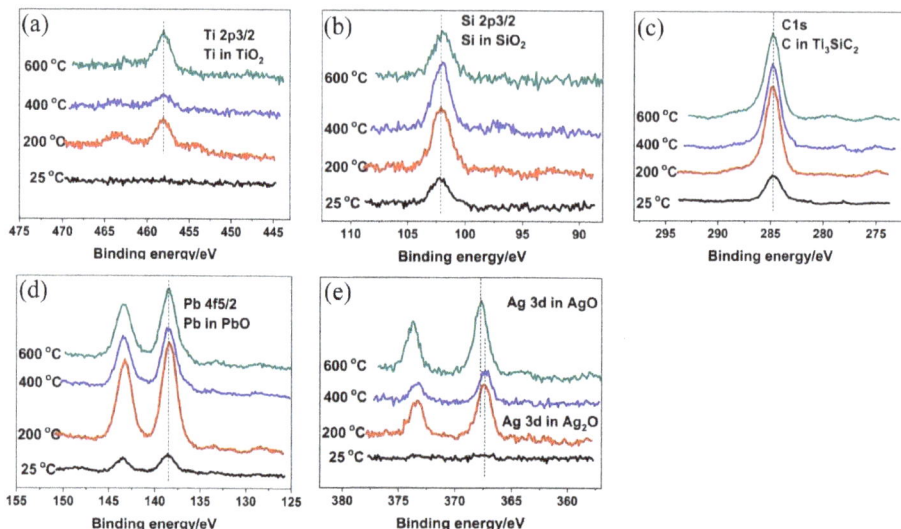

Figure 7. X-ray photoelectron spectroscopy for (**a**) Ti2p, (**b**) Si2pc, (**c**) Pb4f, (**d**) Ag3d, and (**e**) C1s on the worn surface of TSC-PA after sliding against Inconel 718 alloy with different temperature.

At 400 °C, obviously, the TSC-PA was covered with some large wear debris after tribology testing (see Figure 6e). It was expected that, initially, a tribo-oxidation film containing TiO_2, SiO_2, PbO and Ag_2O was formed on its surface (as evident owing to the XPS at 400 °C in Figure 7), which was similar to the oxidation composition at 200 °C. Then, the peel-off and removal of the tribo-oxidation film occurred for TSC-PA. As the sliding finished, the mutual material transfer between the pin and disk was detected by EDS analysis (see Table 3). It was seen from Figure 6f that a relatively smooth film was built up on the surface of the Inconel 718. The plastic flow, in some way, retarded the abrasive wear of TSC-PA; therefore, the wear rate of TSC-PA at 400 °C exhibited a distinct decline in contrast to those at RT and 200 °C. Moreover, the addition of Ag and Pb could strengthen the ductility of TSC-PA. On other hand, when the asperities of the tribo-pair were in contact with each other, the instant flash temperature may reach 3000 °C [32]. Therefore, a high temperature was beneficial for the formation of oxides, such as TiO_2, SiO_2, PbO, and Ag_2O. Therefore, the adhesive wear and plastic flow was the main wear mechanism with slight abrasive wear (as distinct on account of the slight wear trenches on the Inconel 718 disk).

At 600 °C, the worn surface of TSC-PA was surrounded by a relatively smooth and continuous tribolayer after tribology testing (see Figure 6g). This surface was covered with a sufficiently thick tribo-oxidation film containing TiO_2, SiO_2, PbO, and AgO (see Figure 7). In comparison with 400 °C, AgO was found on the worn surface of TSC-PA at 600 °C instead of Ag_2O. Interestingly, the existence of AgO played a significant role in the decrease of friction and wear of the tribopair. The worn surface of Inconel 718 was embraced by a tribo-oxidation film, which was transferred from TSC-PA (see Figure 6h). The serious plastic flowing resulted in the adhesive wear. Mutual transfer of matter between the pin and the disk was verified by the EDS results in Table 3. Moreover, with the increase in temperature, especially at above 400 °C, oxygen contents of the pin and the disk were higher, which indicated the thicker thickness of the tribo-oxidation film (see Table 3). Therefore, the friction coefficient of TSC-PA was lower and more stable and its wear rate was low due to the moderately thick oxides. The wear mechanism was adhesive wear at this temperature.

At 800 °C, the worn surface of TSC-PA was surrounded by a tribo-oxidation film, which was smoothest and densest among all temperatures (see Figure 6). With the thickness of the tribofilm increasing, the covering of the Inconel 718 surface increased. Matter accumulation on the Inconel 718 led to seizure of the contact. The tribofilm reached a critical thickness so that it was easily peeled off from TSC-PA (see Figure 6i). The Inconel 718 showed a smooth tribofilm with large negative wear (see Figures 5b and 6j). Of course, mutual transfer of matter between the pin and the disk was also confirmed by the EDS data in Table 3. The wear mechanism was adhesive wear and tribochemistry of elements of Pb, Ag, Ti, and Si to form TiO_2, SiO_2, PbO, and AgO (see Figure 7).

Conclusively at RT and 200 °C, the abrasive wear ascribed to the peeling-off and pulled-out of Ti_3SiC_2 grains dominated the wear mechanism of TSC-PA, leading to the indispensable friction and wear. As the temperature increased, the formation of tribo-oxidation films was favorable for the reduction in friction and wear of the tribopair. As for the reason why the Inconel 718 disk showed negative wear at elevated temperatures. We proposed that with the temperature increasing, especially at 800 °C, the formed oxidation film was so thick that it was easily peeled-off from the TSC-PA pin and transferred to the Inconel 718 disk, leading to the negative wear of the disk. Additionally, the transition of wear mechanism from abrasive wear to adhesive wear occurred with the increasing temperature. The co-addition of PbO and Ag worked together in the improvement of the tribological behavior of TSC-PA at elevated temperatures, which shed light on the effectiveness of co-addition of solid lubricants in MAX phases at elevated temperatures.

4. Conclusions

The Ti_3SiC_2-Pb-Ag self-lubricating composite (TSC-PA) was successfully sintered by spark plasma sintering technique in vacuum, and its friction and wear behavior were investigated by sliding it against Inconel 718 using a pin-on disk setup at RT and elevated temperatures in air. The results found that the TSC-PA was mainly composed of Ti_3SiC_2, Pb, and Ag. The mechanical properties of the TSC-PA were relatively weaker than that of TSC monolithic. As the temperature increased, the average friction coefficient of TSC-PA/Inconel 718 decreased. The wear rate of TSC-PA exhibited a downward trend and that of Inconel 718 decreased, too, with the rise in the temperature. Interestingly, the wear rate of Inconel 718 was positive wear only at RT and changed to negative wear at other temperatures. The wear mechanism was abrasive wear at low and medium temperatures and turned to adhesive wear at higher and elevated temperatures. The tribochemistry of the sliding contact plays an important part in the transition of the wear theories. It was an effective method of the co-addition of solid lubricants in MAX phases.

Author Contributions: Conceptualization, R.Z.; Data curation, H.Z.; Formal analysis, R.Z. and F.L.; Investigation, R.Z.; Methodology, R.Z. and H.Z.; Writing—original draft, R.Z. and F.L.; Writing—review and editing, R.Z. and F.L. All authors have read and agreed to the published version of the manuscript.

Funding: This research received no external funding.

Institutional Review Board Statement: Not applicable.

Informed Consent Statement: Not applicable.

Data Availability Statement: The data presented in this study are available within the article.

Acknowledgments: The author acknowledges the financial support from the "Xinjiang Tianchi hundred Talents Program (Zhang Rui)".

Conflicts of Interest: The authors declare no conflict of interest.

References

1. Barsoum, M.W. The $M_{N+1}AX_N$ phases: A new class of solids: Thermodynamically stable nanolaminates. *Prog. Solid State Chem.* **2000**, *28*, 201–281. [CrossRef]
2. Gupta, S.; Filimonov, D.; Zaitsev, V.; Palanisamy, T.; Barsoum, M. Ambient and 550°C tribological behavior of select MAX phases against Ni-based superalloys. *Wear* **2008**, *264*, 270–278. [CrossRef]
3. Gupta, S.; Barsoum, M. On the tribology of the MAX phases and their composites during dry sliding: A review. *Wear* **2011**, *271*, 1878–1894. [CrossRef]

4. El-Raghy, T.; Blau, P.; Barsoum, M.W. Effect of grain size on friction and wear behavior of Ti_3SiC_2. *Wear* **2000**, *238*, 125–130. [CrossRef]
5. Hu, C.; Zhou, Y.; Bao, Y.; Wan, D. Tribological Properties of Polycrystalline Ti_3SiC_2 and Al_2O_3-Reinforced Ti_3SiC_2 Composites. *J. Am. Ceram. Soc.* **2006**, *89*, 3456–3461. [CrossRef]
6. Ren, S.; Meng, J.; Lu, J.; Yang, S. Tribological Behavior of Ti_3SiC_2 Sliding Against Ni-based Alloys at Elevated Temperatures. *Tribol. Lett.* **2008**, *31*, 129–137. [CrossRef]
7. Souchet, A.; Fontaine, J.; Belin, M.; Le Mogne, T.; Loubet, J.-L.; Barsoum, M.W. Tribological duality of Ti_3SiC_2. *Tribol. Lett.* **2005**, *18*, 341–352. [CrossRef]
8. Zhang, Y.; Ding, G.P.; Zhou, Y.C.; Cai, B.C. Ti_3SiC_2-a self-lubricating ceramic. *Mater. Lett.* **2002**, *55*, 285–289. [CrossRef]
9. Zhai, H.; Huang, Z.; Zhou, Y.; Zhang, Z.; Wang, Y.; Ai, M. Oxidation layer in sliding friction surface of high-purity Ti_3SiC_2. *J. Mater. Sci.* **2004**, *39*, 6635–6637. [CrossRef]
10. Ren, S.; Meng, J.; Lu, J.; Yang, S.; Wang, J. Tribo-physical and tribo-chemical aspects of WC-based cermet/Ti_3SiC_2 tribo-pair at elevated temperatures. *Tribol. Int.* **2010**, *43*, 512–517. [CrossRef]
11. Dang, W.; Ren, S.; Zhou, J.; Yu, Y.; Wang, L. The tribological properties of Ti_3SiC_2/Cu/Al/SiC composite at elevated temperatures. *Tribol. Int.* **2016**, *104*, 294–302. [CrossRef]
12. Yang, D.; Zhou, Y.; Yan, X.; Wang, H.; Zhou, X. Highly conductive wear resistant Cu/Ti_3SiC_2(TiC/SiC) co-continuous composites via vacuum infiltration process. *J. Adv. Ceram.* **2020**, *9*, 83–93. [CrossRef]
13. Yang, J.; Gu, W.; Pan, L.M.; Song, K.; Chen, X.; Qiu, T. Friction and wear properties of in situ (TiB_2+TiC)/Ti_3SiC_2 composites. *Wear* **2011**, *271*, 2940–2946. [CrossRef]
14. Zhang, R.; Feng, W.; Liu, F. Tribo-oxide Competition and Oxide Layer Formation of Ti_3SiC_2/CaF_2 Self-Lubricating Composites during the Friction Process in a Wide Temperature Range. *Materials* **2021**, *14*, 7466. [CrossRef] [PubMed]
15. Islak, B.Y.; Candar, D. Synthesis and properties of TiB_2/Ti_3SiC_2 composites. *Ceram. Int.* **2021**, *47*, 1439–1446. [CrossRef]
16. Magnus, C.; Cooper, D.; Ma, L.; Rainforth, W.M. Microstructures and intrinsic lubricity of in situ Ti_3SiC_2–$TiSi_2$–TiC MAX phase composite fabricated by reactive spark plasma sintering (SPS). *Wear* **2019**, *448–449*, 203169. [CrossRef]
17. Yang, S.H.; Kong, H.; Yoon, E.-S.; Kim, D.E. A wear map of bearing steel lubricated by silver films. *Wear* **2003**, *255*, 883–892. [CrossRef]
18. Akbulut, M.; Alig, A.R.G.; Israelachvili, J. Friction and tribochemical reactions occurring at shearing interfaces of nanothin silver films on various substrates. *J. Chem. Phys.* **2006**, *124*, 174703. [CrossRef]
19. AlAnazi, F.; Ghosh, S.; Dunnigan, R.; Gupta, S. Synthesis and tribological behavior of novel Ag- and Bi-based composites reinforced with Ti_3SiC_2. *Wear* **2017**, *376–377*, 1074–1083. [CrossRef]
20. Zeng, J.L.; Hai, W.X.; Meng, J.H.; Lu, J.J. Friction and wear of Ti_3SiC_2-Ag/Inconel 718 tribo-pair under a hemisphere-on-disk contact. *Key Eng. Mater.* **2014**, *602–603*, 507–510. [CrossRef]
21. Zhang, R.; Feng, K.; Meng, J.; Su, B.; Ren, S.; Hai, W. Synthesis and characterization of spark plasma sintered Ti_3SiC_2/Pb composites. *Ceram. Int.* **2015**, *41*, 10380–10386. [CrossRef]
22. Zhang, R.; Feng, K.; Meng, J.; Liu, F.; Ren, S.; Hai, W.; Zhang, A. Tribological behavior of Ti_3SiC_2 and Ti_3SiC_2/Pb composites sliding against Ni-based alloys at elevated temperatures. *Ceram. Int.* **2016**, *42*, 7107–7117. [CrossRef]
23. Lu, J.R.; Zhou, Y.; Zheng, Y.; Li, H.Y.; Li, S.B. Interface structure and wetting behaviour of Cu/Ti_3SiC_2 system. *Adv. Appl. Ceram.* **2014**, *114*, 39–44. [CrossRef]
24. Zhou, Y.; Gu, W. Chemical reaction and stability of Ti_3SiC_2 in Cu during high-temperature processing of Cu/Ti_3SiC_2 composites. *Z. Metallkd.* **2004**, *95*, 50–56. [CrossRef]
25. Dang, W.; Ren, S.; Zhou, J.; Yu, Y.; Li, Z.; Wang, L. Influence of Cu on the mechanical and tribological properties of Ti_3SiC_2. *Ceram. Int.* **2016**, *42*, 9972–9980. [CrossRef]
26. Yang, S.; Sun, Z.; Yang, Q.; Hashimoto, H. Effect of Al addition on the synthesis of Ti_3SiC_2 bulk material by pulse discharge sintering process. *J. Eur. Ceram. Soc.* **2007**, *27*, 4807–4812. [CrossRef]
27. El-Raghy, T.; Barsoum, M.W.; Sika, M. Reaction of Al with Ti_3SiC_2 in the 800–1000 °C temperature range. *Mater. Sci. Eng. A* **2001**, *298*, 174–178. [CrossRef]
28. Tzenov, N.; Barsoum, M.; El-Raghy, T. Influence of small amounts of Fe and V on the synthesis and stability of Ti_3SiC_2. *J. Eur. Ceram. Soc.* **2000**, *20*, 801–806. [CrossRef]
29. Zhang, Z.F.; Sun, Z.M.; Zhang, H.; Hashimoto, H. Micron-scale deformation and damage mechanisms of Ti_3SiC_2 crystals induced by indentation. *Adv. Eng. Mater.* **2004**, *6*, 980–983. [CrossRef]
30. El-Raghy, T.; Zavaliangos, A.; Barsoum, M.W.; Kalidindi, S.R. Damage Mechanisms around Hardness Indentations in Ti_3SiC_2. *J. Am. Ceram. Soc.* **2005**, *80*, 513–516. [CrossRef]
31. Woydt, M.; Skopp, A.; Dörfel, I.; Witke, K. Wear engineering oxides/anti-wear oxides. *Wear* **1998**, *218*, 84–95. [CrossRef]
32. Kalin, M. Influence of flash temperatures on the tribological behaviour in low-speed sliding: A review. *Mater. Sci. Eng. A* **2004**, *374*, 390–397. [CrossRef]

Article

Effect of Al$_2$TiO$_5$ Content and Sintering Temperature on the Microstructure and Residual Stress of Al$_2$O$_3$–Al$_2$TiO$_5$ Ceramic Composites

Kunyang Fan [1,2,*], Wenhuang Jiang [1,2], Jesús Ruiz-Hervias [3], Carmen Baudín [4], Wei Feng [1,2], Haibin Zhou [5], Salvador Bueno [6] and Pingping Yao [5]

1. School of Mechanical Engineering, Chengdu University, Chengdu 610106, China; J15983027227@163.com (W.J.); fengwei@cdu.edu.cn (W.F.)
2. Sichuan Province Engineering Technology Research Center of Powder Metallurgy, Chengdu University, Chengdu 610106, China
3. Materials Science Department, Universidad Politécnica de Madrid, E.T.S.I. Caminos, Canales y Puertos, C/Profesor Aranguren s/n, 28040 Madrid, Spain; jesus.ruiz@upm.es
4. Instituto de Cerámica y Vidrio, Consejo Superior de Investigaciones Científicas (CSIC), Kelsen 5, 28049 Madrid, Spain; cbaudin@icv.csic.es
5. State Key Laboratory of Powder Metallurgy, Central South University, Changsha 410083, China; zhbtc22@126.com (H.Z.); yaopingpingxx@sohu.com (P.Y.)
6. Department of Chemical, Environmental and Materials Engineering, Campus Las Lagunillas, Universidad de Jaén, s/n, 23071 Jaen, Spain; jsbueno@ujaen.es
* Correspondence: fankunyang@cdu.edu.cn; Tel.: +86-028-8461-6169

Citation: Fan, K.; Jiang, W.; Ruiz-Hervias, J.; Baudín, C.; Feng, W.; Zhou, H.; Bueno, S.; Yao, P. Effect of Al$_2$TiO$_5$ Content and Sintering Temperature on the Microstructure and Residual Stress of Al$_2$O$_3$–Al$_2$TiO$_5$ Ceramic Composites. *Materials* **2021**, *14*, 7624. https://doi.org/10.3390/ma14247624

Academic Editors: Donatella Giuranno and Christos G. Aneziris

Received: 7 October 2021
Accepted: 8 December 2021
Published: 11 December 2021

Publisher's Note: MDPI stays neutral with regard to jurisdictional claims in published maps and institutional affiliations.

Copyright: © 2021 by the authors. Licensee MDPI, Basel, Switzerland. This article is an open access article distributed under the terms and conditions of the Creative Commons Attribution (CC BY) license (https://creativecommons.org/licenses/by/4.0/).

Abstract: A series of Al$_2$O$_3$–Al$_2$TiO$_5$ ceramic composites with different Al$_2$TiO$_5$ contents (10 and 40 vol.%) fabricated at different sintering temperatures (1450 and 1550 °C) was studied in the present work. The microstructure, crystallite structure, and through-thickness residual stress of these composites were investigated by scanning electron microscopy, X-ray diffraction, time-of-flight neutron diffraction, and Rietveld analysis. Lattice parameter variations and individual peak shifts were analyzed to calculate the mean phase stresses in the Al$_2$O$_3$ matrix and Al$_2$TiO$_5$ particulates as well as the peak-specific residual stresses for different *hkl* reflections of each phase. The results showed that the microstructure of the composites was affected by the Al$_2$TiO$_5$ content and sintering temperature. Moreover, as the Al$_2$TiO$_5$ grain size increased, microcracking occurred, resulting in decreased flexure strength. The sintering temperatures at 1450 and 1550 °C ensured the complete formation of Al$_2$TiO$_5$ during the reaction sintering and the subsequent cooling of Al$_2$O$_3$–Al$_2$TiO$_5$ composites. Some decomposition of AT occurred at the sintering temperature of 1550 °C. The mean phase residual stresses in Al$_2$TiO$_5$ particulates are tensile, and those in the Al$_2$O$_3$ matrix are compressive, with virtually flat through-thickness residual stress profiles in bulk samples. Owing to the thermal expansion anisotropy in the individual phase, the sign and magnitude of peak-specific residual stress values highly depend on individual *hkl* reflection. Both mean phase and peak-specific residual stresses were found to be dependent on the Al$_2$TiO$_5$ content and sintering temperature of Al$_2$O$_3$–Al$_2$TiO$_5$ composites, since the different developed microstructures can produce stress-relief microcracks. The present work is beneficial for developing Al$_2$O$_3$–Al$_2$TiO$_5$ composites with controlled microstructure and residual stress, which are crucial for achieving the desired thermal and mechanical properties.

Keywords: Al$_2$TiO$_5$; ceramics; crystal structure; residual stresses; neutron diffraction

1. Introduction

Aluminum titanate (AT (Al$_2$TiO$_5$)) is a compound with low thermal expansion, excellent thermal shock resistance, and low thermal conductivity. In view of these properties, it is a suitable second-phase candidate for improving the thermal and mechanical properties of alumina (Al_2O_3)-based ceramic systems [1,2]. Alumina–AT (A–AT) ceramics have attracted considerable attention as flaw-tolerant ceramics in thermal and structural applications, such as thermal insulation liners, diesel particulate filters, vehicle emissions control, and high-temperature flue gas filtration supports, because of their improved flaw tolerance, toughness, and superior thermal properties [3–6]. In addition to the typical investigation of mechanical properties, such as flaw tolerance, crack resistance, and thermal shock resistance, many studies have focused on the reinforcement mechanisms of A–AT ceramics [7–9]. The key factors presumed to be responsible for the toughness and improved flaw tolerance of ceramic composites are residual stresses [7].

High tensile and/or compressive residual stresses are expected to develop in alumina-based composites owing to the thermal expansion mismatch among different phases and thermal expansion anisotropy in each phase during cooling after sintering [10]. In the micro-mechanical interaction between matrix and reinforcing particles, residual stresses appear to perform an important function in the mechanical properties of A–AT composites. Padture et al. [11] and Asmi et al. [12] showed that the addition of AT to alumina could improve flaw tolerance and crack-growth resistance. They found that this resulted from the existence of residual stresses arising from the thermal mismatch between alumina and AT. Skala et al. [7] reported that the residual stresses in A–AT composites are responsible for a wide variety of possible toughening mechanisms (e.g., crack deflection, crack bridging, and microcracking). Residual stresses could influence the capability of materials to absorb energy from external loading and distribute damage, consequently leading to enhanced quasi-plasticity and flaw tolerance. The occurrence of spontaneous microcracking in AT was also reported to be caused by residual stresses induced by the strong thermal expansion anisotropy of AT ($\alpha_{a,AT25-1000\,°C} = -2.4 \times 10^{-6}\,°C^{-1}$, $\alpha_{b,AT25-1000\,°C} = 11.9 \times 10^{-6}\,°C^{-1}$ and $\alpha_{c,AT25-1000\,°C} = 20.8 \times 10^{-6}\,°C^{-1}$) [13], resulting in a low elastic modulus, approximately 25–40 GPa [14]. In contrast, in Botero's study [9], a significantly improved elastic modulus was attained in fine-grained AT using nanoindentation in dense A–AT composites without microcracks; this was attributed to the residual stress interaction between matrix and reinforcing particles. These statements show that the mechanical properties of A–AT composites are considerably related to residual stresses and microstructure. To improve A–AT ceramic properties, the strict control of the microstructure and clear comprehension of the nature as well as the magnitude of residual stresses generated during fabrication are necessary.

The interest in developing laminated ceramics to achieve superior flaw tolerance is growing [15]. In addition to the effect of laminate stacking design (e.g., layer thickness and stacking structure), detailed residual stress information (e.g., residual stresses between phases and grains) regarding each layer is crucial. This is extremely important to understand the reinforcing mechanism and optimize the resulting residual stress field, which is useful to produce optimum laminated structures with operative reinforcing layers. It was supposed that the residual stresses in ceramics could be modified by selecting the appropriate material design and fabrication process [16,17]. For A–AT ceramics, the addition of AT and the implementation of sintering treatments can considerably influence their microstructure and mechanical properties [18,19]. Because the co-sintering of layers with different compositions in laminated materials is required, the effects of introducing AT and sintering treatments on residual stress development in addition to those related to laminate stacking design must be considered.

Several studies on the residual stress analysis of A–AT composites have been conducted. Singh [20] and Skala et al. [7] used synchrotron radiation diffraction and X-ray diffraction (XRD) techniques in their investigation. They employed single-peak shift analysis for residual strain measurement in functionally graded A–AT composites. However, they only considered the shifts of individual peaks of Al_2O_3 and AT; residual strain was presented without residual stresses. Fluorescence piezo-spectroscopy was implemented to detect stress in each layer of the multilayered Al_2O_3–Al_2TiO_5 composites. The results indicated weak tension in AT layers (5–20 MPa) and compressive hydrostatic stresses in alumina layers (20 MPa) [10]. The residual stresses in each A–AT laminate layer were also evaluated using simplified model calculations [21]. The reported tensile stresses in A-10AT (composite containing 10 vol.% AT) and compressive stresses in A-30AT (containing 30 vol.% AT) layers were expected to be approximately 15 and 90 MPa, respectively. According to previous work, the residual stress results were presented in different values, depending on fabrication routes, composition design, and residual stress determination techniques. Most studies focus on uniform residual stresses in between layers, laminates, or phases. However, they do not account for thermal and elastic crystal anisotropies at the grain-scale level, which may perform a significant function in crack initiation processes, such as fatigue or failure. Furthermore, owing to crystal anisotropy and microstructure complexity, the residual strains from single-peak diffraction measurements or model calculations may not represent bulk material behavior [22]. To date, direct and reliable experimental data on multi-scale residual stress in A–AT composites remain lacking. To accurately quantify residual stresses under different fabrication conditions or material compositions, the most suitable residual stress measurement technique for each case may be implemented.

The time-of-flight (TOF) neutron diffraction technique allows for the non-destructive measurement of residual strains in bulk materials because it is capable of high neutron penetration. In addition, the entire diffraction pattern, which is important for determining the residual stress in complex materials, such as ceramic composites, can be determined [23]. Information on multi-scale residual stresses, such as average stress among phases, intergranular stress state within a single phase, and even nonuniform microstrains at the atomic scale, can be derived from the analysis of the entire diffraction pattern [22].

In this study, TOF neutron diffraction and Rietveld analysis were applied to precisely determine the residual stresses in a series of A–AT composites with different AT contents and sintering treatments and, therefore, with significantly different microstructures. Multiphase qualitative and quantitative analyses as well as crystal structure determination were performed using Rietveld refinement. The through-thickness residual stress profiles of the mean phase stress and intergranular stress state of each phase were obtained for all the examined A–AT composites. The effects of second-phase AT addition and sintering temperature were discussed based on the observed microstructures and detected residual stresses.

2. Experimental

2.1. Sample Preparation and Characterization

High-purity α-Al_2O_3 (Condea HPA 0.5, USA; d_{50} = 0.35 µm) and TiO_2 (anatase; Merck, 808, Darmstadt, Germany; d_{50} = 0.35 µm) powders were used as starting powders. Monoliths of Al_2O_3–Al_2TiO_5 (A–AT) composites with 10 and 40 vol.% of AT were manufactured by the reaction sintering of green compacts prepared from mixtures of Al_2O_3 and TiO_2 with relative TiO_2 contents of 5 and 20 wt.%, respectively. Following colloidal filtration techniques [24], the powder mixtures were dispersed in deionized water by adding 0.5 wt.% (on a dry solids basis) of a carbonic acid-based polyelectrolyte (Dolapix CE64, Zschimmer-Schwarz, Lahnstein, Germany). Stable suspensions of a mixture of undoped alumina and titania with a solid loading of 80 wt.% were prepared by 4 h ball milling, using an alumina jar and balls. Green compact blocks, 70 mm × 70 mm with 12 mm thickness, were slip cast in plaster of Paris molds. The cast bodies were carefully removed from the molds and dried in air at room temperature for a minimum of 24 h. The dried blocks were sintered in an electrical box furnace (Termiber, Madrid, Spain) with heating and cooling rates of 2 °C/min. The first step involved sintering at 1200 °C with a 4 h dwell time during heating. In the second step, two different treatments with different maximum sintering temperatures, i.e., 2 h dwell times at 1450 and 1550 °C, were implemented. The materials were denoted as A-10AT(1450), A-10AT(1550), A-40AT(1450), and A-40AT(1550) to describe compositions and sintering temperatures.

The phase identification of the sintered specimens was performed using an X-ray diffractometer (XRD, DX-2700, Dandong, China) using CuKα radiation at room temperature over a 2θ range of 10 to 80°, with a step size of 0.03° and a counting time of 1.0 s at each step. Quantitative phase analysis was undertaken using the Rietveld pattern fitting method. Microstructural characterization of polished and chemically etched (HF 10 vol.%, 1 min) samples was performed by scanning electron microscopy (SEM). The average grain sizes of Al_2O_3 and AT particles were determined using the linear intercept method, considering at least 200 grains for each phase. The densities of samples were determined using Archimedes' method (European Standard EN 1389:2003). The relative densities of α-Al_2O_3 (ASTM 42-1468) and AT (ASTM 26-0040) were calculated as percentages of their theoretical densities, i.e., 3.99 and 3.70 g/cm^3, respectively. The flexural strength was determined by a three-point bending test on the rectangular bar samples (geometry of 28 × 10 × 2 mm^3), under conditions of a span length of 20 mm and a constant loading speed of 0.5 mm/min. For each material, three samples were prepared for the bending test to obtain an average value and the standard deviation. The fracture surfaces of the fractured samples were characterized by SEM after the three-point bending test.

2.2. Residual Stress Measurement

For neutron diffraction strain scanning, rectangular parallelepiped samples (30 × 30 × 10 mm^3) were employed. Residual strain measurements were performed by neutron diffraction on the ENGIN-X TOF instrument [25] at the ISIS neutron and muon source in Rutherford Laboratory, UK. Two detector banks were set at Bragg angles of $2\theta_B = \pm 90°$ for simultaneous strain measurement in two directions: the in-plane direction (parallel to the major plane of 30 mm × 30 mm samples) and normal direction (perpendicular to the major plane of 30 mm × 30 mm samples). The experimental setup is illustrated in Figure 1.

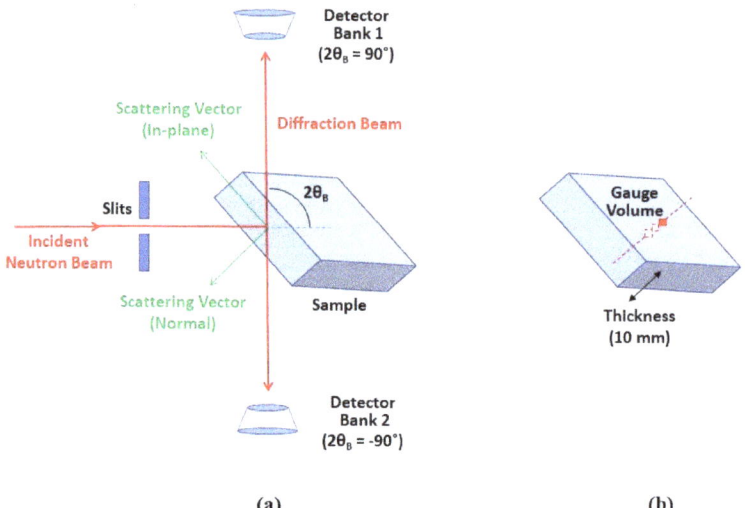

Figure 1. Experimental setup on TOF neutron strain scanner ENGIN-X at ISIS. (**a**) Residual strains were collected both in in-plane and normal directions. (**b**) Strain scanning was implemented along sample thickness with interior gauge volume.

The set measurement gauge volume was 16 × 3 × 1 mm^3, with the centroid defined as the location of measurement. Through-thickness strain scanning was implemented along the sample thickness (10 mm) with a 1.5 mm measurement step. To avoid anomalous strain due to the effects of partial neutron gauge volume filling near the surface [26], the scanning points were no closer than 2 mm from the sample surfaces. The stress-free reference lattice parameters of Al_2O_3 and AT phases were obtained by measuring Al_2O_3 and AT powders with the same gauge volume. As the experimental standard, CeO_2 powder was measured to calibrate the peak profile parameters to be employed in Rietveld refinement. All the measurements were performed at room temperature.

2.3. Data Analysis

The entire diffraction spectrum was obtained from TOF diffraction measurements. Diffraction spectra were analyzed by the Rietveld refinement of the entire spectrum and by the single-peak fitting of each hkl reflection using the TOPAS-Academic V5 software package [27].

2.3.1. Rietveld Refinement

Rietveld refinement is a well-known structure refinement method of the entire profile based on least-squares fitting. It is powerful for determining the complete crystal structure (e.g., lattice parameters, atomic coordinates, and occupancy) of a multiphase material [28,29]. The peak position, intensity, and width of each phase were independently derived. Quantitative phase analysis was also performed using Rietveld analysis.

The ENGIN-X instrument diffraction profile was modeled using a convolution of a pseudo-Voigt function, $pV(t)$, with two back-to-back exponentials, $E(t)$ [22]. Instrument-dependent diffractometer constants and profile parameters were calibrated by the refinement of data from the CeO_2 standard powder (cubic phase, space group $Fm\bar{3}m$, and a = 5.4114 Å). The derived parameters were then used to refine the studied bulk samples and stress-free reference powders.

The refinements of A–AT composites were undertaken with a two-phase model consisting of a hexagonal α-Al_2O_3 phase and an orthorhombic AT phase. Parameters, such as lattice parameters, atomic coordinates, isotropic thermal parameters, scale factors, and polynomial background parameters, were improved by Rietveld refinement. All refinements were implemented step-by-step to avoid correlation effects among parameters. The space groups and initial atomic structure information used in the refinements are listed in Table 1. The overall fitting quality was assessed in terms of R values [30] obtained from the refinements.

Table 1. Initial crystal structure parameters of Al_2O_3 and Al_2TiO_5 phase used in refinement.

Phase	Space Group	Lattice Parameter/Å	Ion Position and Coordinate		
			Ion	x	y	z
α-Al_2O_3 [31] (hexagonal)	$R\bar{3}c$	$a = b$ = 4.7602 c = 12.9933 $\alpha = \beta = 90°$ $\gamma = 120°$	Al^{3+} O^{2-}	0 0.30624	0 0	0.35216 0.25
Al_2TiO_5 [13] (orthorhombic)	$Cmcm$	c = 3.593 a = 9.433 b = 9.641 $\alpha = \beta = \gamma = 90°$	Ti_1^{4+} Al_1^{3+} Ti_2^{4+} Al_2^{3+} O_1^{2-} O_2^{2-} O_3^{2-}	0 0 0 0 0 0 0	0.1854 0.1854 0.13478 0.13478 0.7577 0.0485 0.3125	0.25 0.25 0.5615 0.5615 0.25 0.1167 0.0721

2.3.2. Stress Determination

Based on the entire diffraction spectrum, residual stresses can be calculated either from the change in lattice parameters of each phase or from individual hkl peak shifts. The former represents the average stress behavior of each phase (called mean phase stress), and the latter represents the stress in individual grains (called peak-specific residual stress).

Mean Phase Stresses

The mean phase strain, given by the weighted average of several single-peak strains, was determined from the change in the average lattice parameters of each phase in composites with respect to those in stress-free reference powders [28]. In this case, all diffraction peaks were considered, thus allowing the representation of residual stresses more precisely and reliably for bulk composites.

In this study, the strain for the hexagonal α-Al$_2$O$_3$ phase can be calculated along lattice axes a and c:

$$\varepsilon_a = \frac{a - a_0}{a_0} \tag{1}$$

$$\varepsilon_c = \frac{c - c_0}{c_0} \tag{2}$$

where a ($a = b$) and c are the lattice parameters of the α-Al$_2$O$_3$ phase, and a_0 and c_0 are the stress-free lattice parameters measured from the Al$_2$O$_3$ starting powder. For orthorhombic AT, the strains in the a, b, and c axes can be similarly determined.

Without considering granular anisotropy, the mean phase strains of the Al$_2$O$_3$ and AT phases in gauge volume were calculated by averaging the strain over the unit cell, as presented in [32].

$$\bar{\varepsilon}_A = \frac{1}{3}(2\varepsilon_a + \varepsilon_c)_A \tag{3}$$

$$\bar{\varepsilon}_{AT} = \frac{1}{3}(\varepsilon_a + \varepsilon_b + \varepsilon_c)_{AT} \tag{4}$$

Considering the fabrication process and geometrical symmetry of samples, the strains were assumed to be isotropic in a direction parallel to the major plane of samples. Thus, the strains measured in the in-plane and normal directions are sufficient for stress determination. These are used as principal strains ε_{ii} (i = 1, 2, 3), $\varepsilon_{11} = \varepsilon_{22} = \varepsilon_{In-plane}$, and $\varepsilon_{33} = \varepsilon_{Normal}$. Thus, the mean phase stresses of Al$_2$O$_3$ and AT were determined in the in-plane and normal directions using Hooke's law, respectively, as follows:

$$\sigma_{In-plane} = \frac{E}{1+\nu}\bar{\varepsilon}_{In-plane} + \frac{E\nu}{(1+\nu)(1-2\nu)}\left(2\bar{\varepsilon}_{In-plane} + \bar{\varepsilon}_{Normal}\right) \tag{5}$$

$$\sigma_{Normal} = \frac{E}{1+\nu}\bar{\varepsilon}_{Normal} + \frac{E\nu}{(1+\nu)(1-2\nu)}\left(2\bar{\varepsilon}_{In-plane} + \bar{\varepsilon}_{Normal}\right) \tag{6}$$

where $\bar{\varepsilon}$ corresponds to the calculated mean phase strains of Al$_2$O$_3$ and AT phases, as given by Equations (3) and (4), respectively; and E and ν are the bulk elastic constants of Al$_2$O$_3$ (E = 400 GPa and ν = 0.22) [33] and AT (E = 284 GPa and ν = 0.33) [9].

Peak-Specific Residual Stresses

Owing to single-crystal anisotropy, the strain measured by diffraction varies among grain families; consequently, such is the case among diffraction peaks [34]. The deviation from the average phase stress in a given grain, that is, the peak-specific residual stresses, can be obtained by measuring the shifts in individual diffraction peaks. Understanding peak-specific residual stresses is important because the stress concentrations at the scale of individual grains may ultimately affect crack initiation processes in fatigue or brittle failure.

In TOF diffraction, at a fixed scattering angle, $2\theta_B$, the lattice spacing, d_{hkl}, of the hkl plane can be obtained from the TOF of the peak position, t_{hkl} [23]:

$$d_{hkl} = \frac{h}{2\sin\theta_B mL} t_{hkl} \tag{7}$$

where h is the Planck constant, m is the neutron mass, and L is the neutron flight path length.

The peak-specific residual strain, which depends on the change in the lattice spacing of the Δd_{hkl} hkl plane, can then be calculated in terms of the TOF shift in the recorded peak, Δt_{hkl}:

$$\varepsilon_{hkl} = \Delta d_{hkl}/d^0_{hkl} = \Delta t_{hkl}/t^0_{hkl} \tag{8}$$

where d^0_{hkl} is the strain-free reference lattice spacing, t^0_{hkl}, and strain-free reference TOF at the hkl peak. The peak positions can be precisely determined by single-peak fitting, with a typical sensitivity of $\delta_\varepsilon = \Delta d_{hkl}/d_{hkl} \cong 50 \times 10^{-6}$. The peak-specific residual strains measured along the in-plane and normal directions were characterized as $\varepsilon^{hkl}_{In-plane}$ and $\varepsilon^{hkl}_{Normal}$, respectively. They were considered as principal strain components ε_{ii} (i = 1, 2, 3; $\varepsilon_{11} = \varepsilon_{22} = \varepsilon_{In-plane}$; and $\varepsilon_{33} = \varepsilon_{Normal}$).

The peak-specific residual stresses for each reflection of Al_2O_3 and AT in the in-plane and normal directions ($\sigma^{hkl}_{In-plane}$ and σ^{hkl}_{Normal}) were calculated using Hooke's law, as follows:

$$\sigma^{hkl}_{In-plane} = \frac{E_{hkl}}{1+v_{hkl}}\varepsilon^{hkl}_{In-plane} + \frac{E_{hkl}v_{hkl}}{(1+v_{hkl})(1-2v_{hkl})}\left(2\varepsilon^{hkl}_{In-plane} + \varepsilon^{hkl}_{Normal}\right) \tag{9}$$

$$\sigma^{hkl}_{Normal} = \frac{E_{hkl}}{1+v_{hkl}}\varepsilon^{hkl}_{Normal} + \frac{E_{hkl}v_{hkl}}{(1+v_{hkl})(1-2v_{hkl})}\left(2\varepsilon^{hkl}_{In-plane} + \varepsilon^{hkl}_{Normal}\right) \tag{10}$$

where E_{hkl} and v_{hkl} are the diffraction elastic constants (DECs) corresponding to the hkl reflection in each corresponding phase. In the present work, the DECs of Al_2O_3 (hkl) and AT (hkl) were obtained using the program IsoDEC [35,36], following the Kröner model.

3. Results and Discussions

3.1. Phase Composition and Microstructure

The XRD profiles of the studied A–AT composites are shown in Figure 2.

In the A-10AT(1450) and A-40AT(1450) composites, only peaks corresponding to the Al_2O_3 and AT phases were detected without the TiO_2 phase. This demonstrates that the initial titania powders completely reacted with alumina and transformed into AT during fabrication. A previous study [37] investigated the dynamic phase formation in the temperature range of 20–1400 °C for the fabrication of A–AT samples by means of neutron diffraction and differential thermal analysis. The results showed that the formation of AT in A–AT samples occurred at temperatures exceeding 1310 °C by the reaction sintering of the Al_2O_3 and TiO_2 mixture; TiO_2 disappeared at 1370 °C. Violini et al. [38] reported that in the two-step reaction sintering of initial powders (alumina and titania), the formation of AT started at 1380 and ended at 1440 °C. In our work, the sintering temperature 1450 °C ensured the complete formation of AT. However, in the XRD profiles of the A-10AT(1550) and A-40AT(1550) composites, in addition to the main peaks of the Al_2O_3 and AT phases, a small peak of TiO_2 (rutile) phase was detected at around 2θ = 7.6° (hkl = 110). This indicated the existence of a TiO_2 (rutile) phase in the 1550 °C sintered A–AT composites. The results of quantitative phase analysis demonstrated that the content of TiO_2 was very limited, with values of 0.5 wt.% and 1.2 wt.% in the A-10AT(1550) and A-40AT(1550) composites, respectively.

According to the previous discussion, the complete reaction from initial titania and alumina to AT can be guaranteed with a sintering temperature higher than 1450 °C. The detected TiO_2 phase in these 1550 °C sintered A–AT composites could be formed due to a partial decomposition of the AT phase during cooling. It is well known that pure synthetized AT is thermally unstable in the temperature range of 800–1280 °C. It tends to decompose, through a eutectoid reaction, into α-Al_2O_3 and TiO_2 (rutile) during cooling from sintering treatments [13,39]. This brings disadvantage for the material, such as reduced thermal shock resistance. It is well accepted that the decomposition of AT is a nucleation-and-growth process. Experimental evidence has suggested that AT can be thermally stabilized by limiting its grain growth [40,41]. The heat treatment temperature plays an important role in the grain growth progress. In our study, with a smaller AT grain size in 1450 °C sintered

A–AT materials, compared with the one in 1550 °C sintered samples, the AT phase exhibited increased thermal stability, i.e., without decomposition.

The characteristic microstructures of the investigated A–AT composites are shown in Figure 3.

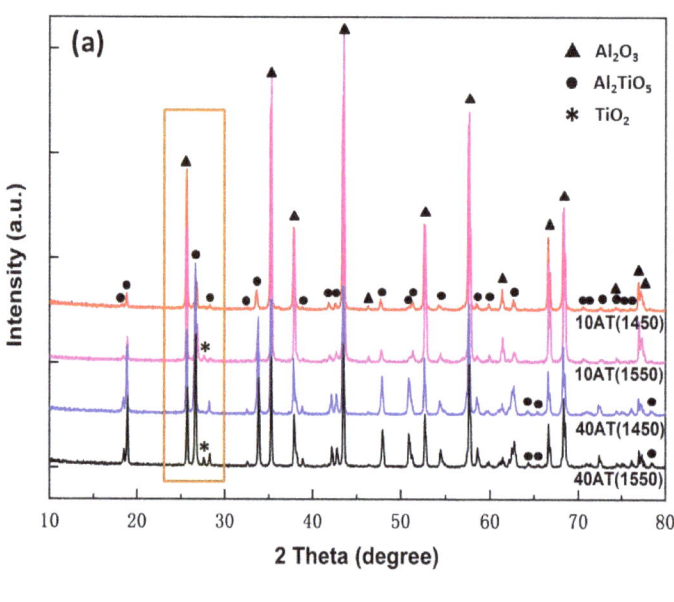

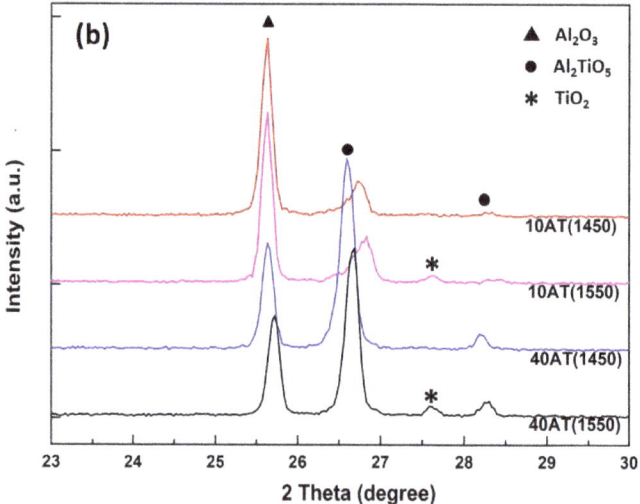

Figure 2. XRD patterns of studied Al_2O_3–Al_2TiO_5 bulk ceramic composites. (**a**) The full XRD patterns at the range of $2\theta = 10\sim80°$ and (**b**) partial enhancement at the range of $2\theta = 23\sim30°$, as marked in (**a**) with a yellow rectangle.

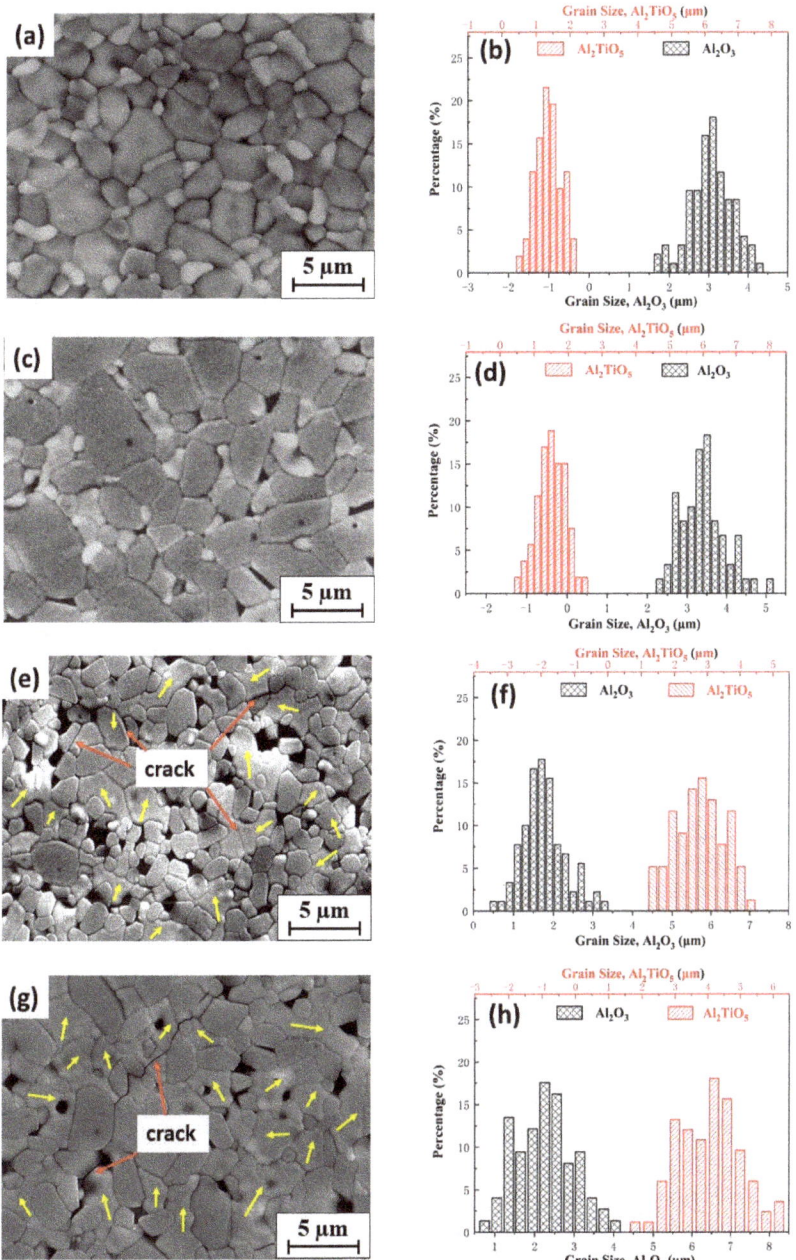

Figure 3. SEM images of polished and etched surfaces of Al_2O_3–Al_2TiO_5 samples and size distribution of Al_2O_3 and Al_2TiO_5 particles in each sample: (**a**,**b**) A-10AT(1450); (**c**,**d**) A-10AT(1550); (**e**,**f**) A-40AT(1450); and (**g**,**h**) A-40AT(1550). Yellow arrows indicate locations of AT phases in A-40AT composites.

A dense microstructure was observed in the A-10AT composites (Figure 3a,c) irrespective of the sintering temperature (i.e., 1450 and 1550 °C). In the back-scattered electron images, AT grains and alumina matrix appeared with light and dark gray colors, respectively. Round-shaped and slightly elongated-shaped AT grains (1.3–1.6 µm) were homogeneously distributed and mainly located at the triple points and grain boundaries of the alumina matrix. A relatively narrow distribution of grain sizes for both the Al_2O_3 matrix and second-phase AT was observed in the A-10AT composites, according to the histogram of grain size distribution in Figure 3b,d. As the AT content increased to 40 vol.%, the grain size of Al_2O_3 decreased, whereas the AT grain size increased (Table 2). The AT grains in the A-40AT composites exhibited a distinct irregular shape and heterogeneity in size distribution, as indicated by the arrows in Figure 3e,g. Some AT grains clustered together as a submatrix (4–6 µm), where alumina grains were separated and surrounded by AT grains; a similar phenomenon was also reported in another study [38]. The Al_2O_3 grain size evidently decreased as the AT content increased owing to the inhibiting effect of second-phase AT particles on the Al_2O_3 matrix grain growth. This phenomenon was also observed in other alumina-based ceramics, such as alumina–zirconia ceramics [22,42]. As the AT content increased, more pores and microcracks were observed in their SEM images. This corresponds to the reduced density tendency summarized in Table 2.

Table 2. Values of measured density (ρ), relative density ($\rho_{relative}$), and average grain size (G) of Al_2O_3 and Al_2TiO_5, and flexural strength (σ_f).

Sample	Al_2TiO_5 (vol.%)	Sintering Temperature (°C)	ρ (g/cm³)	$\rho_{relative}$ (T.D.%)	G (µm) G_A	G_{AT}	σ_f (MPa)
A-10AT(1450)	10	1450	3.86 ± 0.03	97.4 ± 0.1	3.1 ± 0.4	1.3 ± 0.4	221 ± 8
A-10AT(1550)	10	1550	3.84 ± 0.03	97.0 ± 0.1	3.4 ± 0.4	1.6 ± 0.4	194 ± 9
A-40AT(1450)	40	1450	3.57 ± 0.06	92.1 ± 0.2	1.7 ± 0.4	3.0 ± 0.7	83 ± 3
A-40AT(1550)	40	1550	3.53 ± 0.08	91.2 ± 0.2	2.3 ± 0.7	4.1 ± 0.9	61 ± 2

Figure 4 exhibits that microcracks were mainly along the grain boundaries between Al_2O_3 and AT; some transgranular cracks, acting as bridges between one alumina grain to another, were present in the AT grains. Such microcracks propagation behaviors reflect the complex stresses in A–AT composites during fabrication.

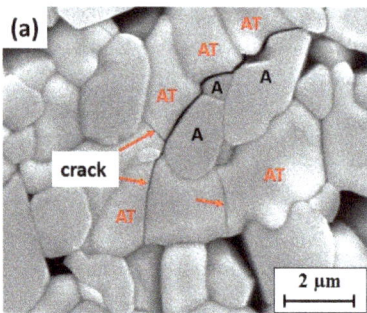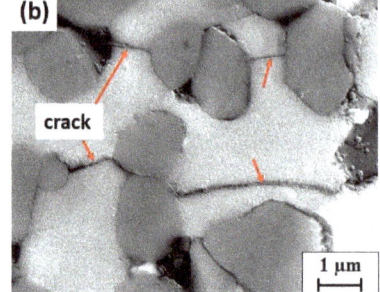

Figure 4. SEM micrograph showing microcracks in A-40AT samples: (**a**) A-40AT(1450). The aluminum titanate and alumina grains are marked as "AT" and "A", respectively; (**b**) BSE-SEM image of A-40AT(1550). Alumina grains in dark gray, and aluminum titanate in light gray. Red arrows indicate locations of microcracks.

The effect of sintering temperature was observed as follows. At higher sintering temperatures with the same composition, the studied A–AT composites presented a coarser microstructure associated with the grain growth of Al_2O_3 and AT. This phenomenon is particularly evident in the A-40AT composites. In these composites, at a higher sintering temperature (i.e., 1550 °C), the excessive grain growth of Al_2O_3 and AT was promoted after the formation of the AT phase. Owing to the crystallographic anisotropy of thermal expansion in the orthorhombic structure, the grain growth of AT particulates became more abnormal. Microcracks form once the grain size of AT reaches the critical size (approximately 2 μm) necessary for microcracking [38,43]. Compared with the A-40AT(1450) composites, more microcracks appeared in the A-40AT(1550) composites.

These microcracks function as excellent regions for main crack energy absorption and dissipation, leading to remarkable crack attenuation and deflection. Such a mechanism can be confirmed by the fracture surface morphologies of A–AT bulk ceramic composites (Figure 5).

 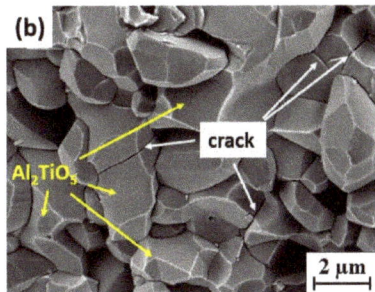

Figure 5. SEM images of fracture surfaces of Al_2O_3–Al_2TiO_5 composites. Arrows indicate Al_2TiO_5 particles and microcracks. (**a**) A-10AT composites irrespective of sintering temperature and (**b**) A-40AT composites irrespective of sintering temperature.

Without microcracking, the predominant fracture mode of A-10AT composites (Figure 5a) was intergranular fracture. In contrast, with weaker grain boundaries and microcracks in the A-40AT composites (Figure 5b), transgranular fracture appeared in AT particles in addition to the intergranular fracture along weak boundaries between particles. The existence of these microcracks and "weak" grain boundaries is presumed to impart low mechanical strength; however, flaw tolerance, such as high thermal shock resistance and improved thermal stability, improves [44]. This point corresponds well with the results of flexure strength of the studied A–AT composites, as given in Table 2. With the increase of AT content, the flexure strength of A–AT composites remarkably decreased, mainly due to the existence of microcracks. As the sintering temperature increased from 1450 to 1550 °C, the flexure strength slightly reduced, which was related to the coarser microstructure and lower density of the composites.

Attempts were made to identify the existence of TiO_2 in the microstructure of 1550 °C sintered A–AT composites. However, no significant evidence of TiO_2 was detected in the SEM micrographs of the A-10AT(1550) and A-40AT(1550) composites, owning to the insufficient content of TiO_2.

3.2. Neutron Diffraction Patterns and Rietveld Analysis

No distinct differences were observed among the TOF diffraction patterns of A–AT samples with the same AT contents at various sintering temperatures; the diffraction spectra mainly varied with the AT content. The representative diffraction patterns of the A-10AT(1550) and A-40AT(1550) composites are shown in Figure 6. As the AT content increased, the peak intensities of the Al_2O_3 matrix decreased; this is in contrast to the increase in AT intensity. In addition, at the same scanning point, the differences between the spectra in the in-plane and normal directions (measured at $2\theta = \pm 90°$) are insignificant.

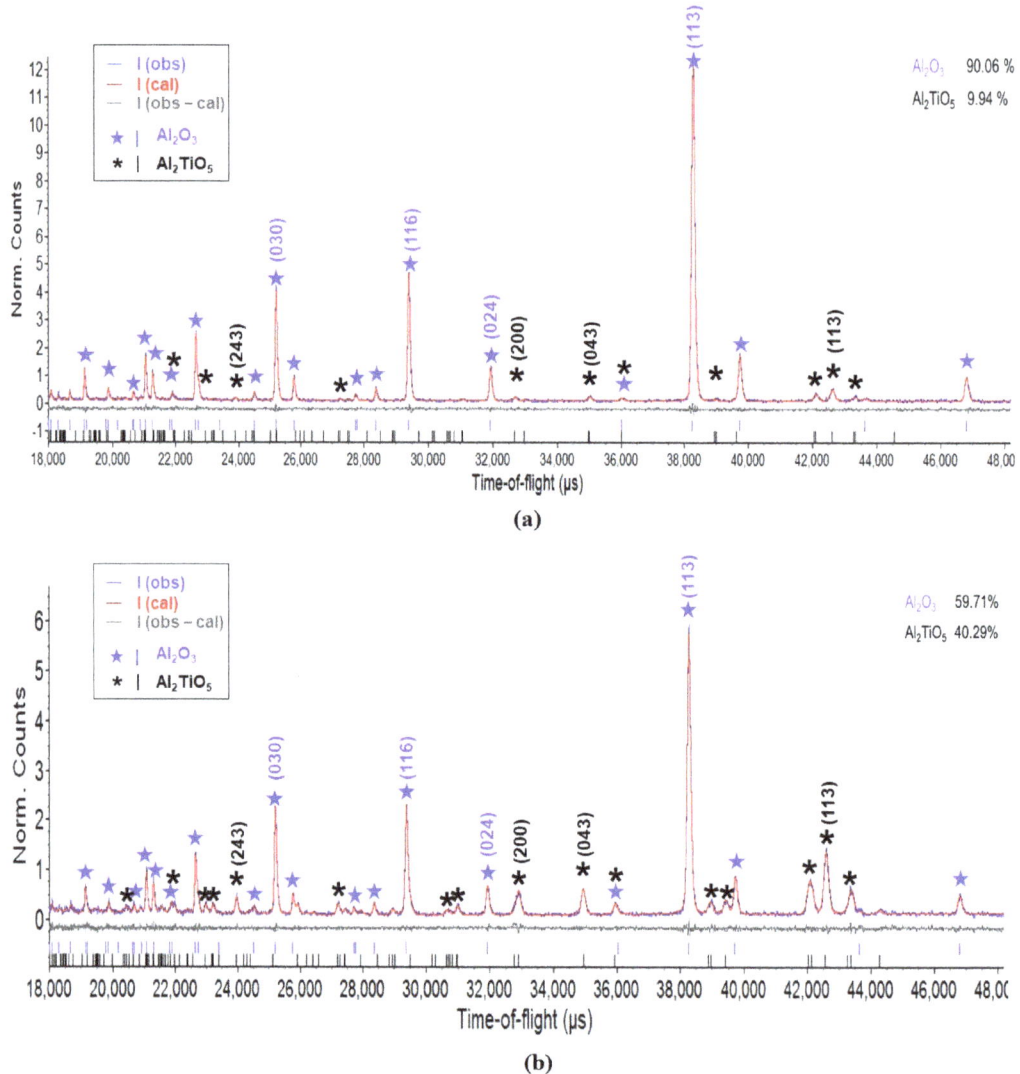

Figure 6. Representative TOF neutron diffraction patterns analyzed by Rietveld method for (**a**) A-10AT(1550) and (**b**) A-40AT(1550) samples. Investigated reflections are marked above peaks with ★ for Al_2O_3 and * for Al_2TiO_5.

In Figure 6, raw data collected by the diffraction method (observed) are represented by a blue line, and data derived by Rietveld refinement are denoted by a red line overlapping with the blue (observed) pattern. Unlike the results in the XRD analysis, only peaks of the Al_2O_3 and AT phases without evidence of the TiO_2 phase are observed in the TOF diffraction patterns of A-AT(1550) samples, which is concerned with the limited content of TiO_2. Due to the difficulty in identifying TiO_2 in the microstructure and neutron diffraction analysis, it can be considered that the effect of TiO_2 on residual stresses of the composites would be extremely limited, and thus the crystal structure and stress analysis of TiO_2 were not involved in the following results and discussion. The Rietveld refinement of a two-phase model consisting of α-Al_2O_3 and AT phases was highly satisfactory for all A–AT bulk samples. The positions of individual peaks of α-Al_2O_3 and AT phases were identified with blue and black tick marks at the bottom of the pattern, respectively. The gray line below the pattern represents the difference between the observed and calculated intensities, which were virtually flat in both the A-10AT and A-40AT composites. This indicates the excellent fits in Rietveld refinement that were achieved for all the reference powders and A–AT bulk samples; the weighted residual error, R_{wp}, ranged from 3 to 7%.

The phase composition corresponding to the measured gauge volumes was evaluated by Rietveld refinement; the volume fraction is shown on the upper right of the diffraction pattern fitting window (Figure 6). Virtually constant AT contents were recorded at different scanning positions of the same sample. The calculated AT volume fractions were 9.5 ± 1 and 40 ± 2 vol.% in the A-10AT and A-40AT composites, respectively; these results agreed well with the nominal values.

3.3. Crystal Structures

After applying Rietveld refinement to TOF neutron diffraction patterns, the crystal structures of Al_2O_3 and AT in the stress-free reference powders and A–AT composites were obtained. Table 3 summarizes the average values of refined lattice parameters of α-Al_2O_3 (a, c) and orthorhombic AT (a, b, and c) at different scanning points in each sample (standard deviations are shown in brackets). Based on these standard deviations, no remarkable variance was found in the lattice parameters of both Al_2O_3 and AT in the in-plane and normal directions as well as at different scanning points in the same bulk sample. This indicates the existence of phases and microstructure distributions that are virtually homogeneous in the above materials.

Table 3. Average lattice parameters (A°) of Al_2O_3 and Al_2TiO_5 phases obtained by Rietveld refinement of initial powders and A–AT composite samples.

Sample	α-Al_2O_3 Phase				Orthorhombic Al_2TiO_5 Phase					
	In-Plane		Normal		In-Plane			Normal		
	a	c	a	c	a	b	c	a	b	c
Al_2O_3 powder	4.76026	12.99488	4.76048	12.99456	-	-	-	-	-	-
Al_2TiO_5 powder	-	-	-	-	3.59348	9.43079	9.63653	3.59375	9.43032	9.63763
A-10AT (1450)	4.7585 (1)	12.9940 (3)	4.7582 (1)	12.9937 (3)	3.5643 (9)	9.4529 (10)	9.7118 (8)	3.5649 (4)	9.4527 (12)	9.7120 (10)
A-10AT (1550)	4.7585 (1)	12.9947 (2)	4.7588 (1)	12.9944 (1)	3.5641 (3)	9.4524 (8)	9.7112 (12)	3.5646 (9)	9.4520 (14)	9.7118 (12)
A-40AT (1450)	4.7588 (1)	12.9920 (2)	4.7591 (1)	12.9920 (4)	3.5863 (3)	9.4385 (9)	9.6534 (6)	3.5867 (4)	9.4388 (5)	9.6540 (10)
A-40AT (1550)	4.7589 (1)	12.9920 (2)	4.7592 (1)	12.9920 (4)	3.5894 (5)	9.4347 (8)	9.6456 (8)	3.5895 (5)	9.4349 (5)	9.6464 (10)

The lattice parameters (a and c) of Al_2O_3 in all the studied A–AT ceramics were found to be slightly lower than those of the initial Al_2O_3 powders. Thus, compressive strain was anticipated in the Al_2O_3 phase of A–AT ceramics. With increasing AT content, changes in the lattice parameters of Al_2O_3 were not remarkable, and the c value only slightly decreased. Compared with the initial AT powders, the a value of orthorhombic AT in all the studied A–AT ceramics decreased, whereas the b and c values increased. As the AT content increased from 10 to 40 vol.%, the a value of AT increased, whereas the b and c values distinctly decreased.

In comparing the A–AT samples with the same composition but fabricated at different sintering temperatures, no remarkable difference was observed among the lattice parameters of the Al_2O_3 and AT crystallites, except for AT in the A-40AT samples. In the A-40AT composites, the materials sintered at a higher temperature value (1550 °C) exhibited an increase in the a values of the AT phase; however, the b and c values reduced compared with the materials sintered at 1450 °C.

These behaviors can be understood by considering the thermal expansion mismatch between the Al_2O_3 and AT phases, the thermal expansion anisotropy of each phase, and the microstructure characteristics of composites. According to the microstructure described above, the A-10AT composites (Figure 3a,b) were dense, and fine AT grains were surrounded by Al_2O_3 matrix grains. The average crystallographic thermal expansion (CTE) coefficient of the Al_2O_3 matrix ($\alpha_{A,25-1000\ °C}$ = 8.6 × 10^{-6} $°C^{-1}$) is smaller than that of the AT inclusions ($\alpha_{AT,25-1000\ °C}$ = 10.1 × 10^{-6} $°C^{-1}$) [24] when subjected to cooling from the maximum sintering temperature to room temperature in the fabrication procedure. Consequently, Al_2O_3 matrix grains were compressed, and lattice parameters a and c decreased.

In contrast, for AT particles, most of the contact and restraints emanated from the Al_2O_3 matrix in the A-10AT composites. Owing to the CTE mismatch between Al_2O_3 and AT, the AT grains were presumed to be in tension, and the lattice parameters increased. However, according to the lattice parameters obtained by neutron diffraction, the b and c values of orthorhombic AT in A-10AT ceramics increased as the a value decreased. The strong CTE anisotropy exhibited by AT ($\alpha_{a,AT25-1000\ °C}$ = −2.4 × 10^{-6} $°C^{-1}$, $\alpha_{b,AT25-1000\ °C}$ = 11.9 × 10^{-6} $°C^{-1}$ and $\alpha_{c,AT25-1000\ °C}$ = 20.8 × 10^{-6} $°C^{-1}$) [13] indicates that this anisotropy is the predominant effect rather than the CTE mismatch with Al_2O_3, although only a small quantity of AT was included in the A–AT ceramics. By increasing the AT content, the contact among AT grains was enhanced. Most restraints were from the closed AT grains, and the effect of CTE anisotropy on the AT phase became more distinct. A decrease in the a value and further increases in b and c values were anticipated in AT. However, the derived lattice parameter results compared with those in the A-10AT ceramics showed that the a value of AT increased in the A-40AT ceramics, whereas the b and c values decreased. Such phenomena can be explained by the microstructure of the A–AT composites. As presented in Figure 3, spontaneous microcracking occurred in the AT of the A-40AT composites. Thus, restraints among grains were evidently released, and changes in the lattice parameters of AT weakened. This also explains the difference in AT lattice parameters between the A-40AT(1450) and A-40AT(1550) ceramics. At higher sintering temperatures, more microcracks were observed in A-40AT(1550) than in A-40AT(1450). Thus, changes in the lattice parameters of AT in the A-40AT(1550) ceramics compared with the A-40AT(1450) materials weakened, leading to an increase in a values of the AT phase but reductions in the b and c values.

In addition, the lattice parameters could reflect the thermal stability of Al_2TiO_5. Skala et al. [13] reported that the thermal stability of Al_2TiO_5 is closely related to its lattice constant c. Its increase will lead to a reduction of the distortion of the octahedra in the crystal structure of Al_2TiO_5, so that the stability of Al_2TiO_5 is improved. In our study, in A–AT ceramics with the same AT addition, as the sintering temperature increased from 1450 °C to 1550 °C, the lattice parameter c of AT reduced, indicating less thermal stability of the AT phase. This explains why the decomposition of AT occurred in the 1550 °C sintered A–AT samples.

3.4. Stress Determination

3.4.1. Mean Phase Stresses

Mean phase stress, the average stress in each phase over several randomly oriented grains within the gauge volume area in the TOF measurement, was calculated using the change in the lattice parameters of the Al_2O_3 and AT phases, as given by Equations (1)–(6), respectively. The through-thickness mean phase stress profiles measured in the in-plane and normal directions are depicted in Figure 7 for the A–AT bulk samples with different AT additions and sintering temperatures (1450 and 1550 °C). Note that error bars corresponding to the statistical uncertainties of the determined lattice parameters are smaller than the size of the symbols employed in the presented graphs.

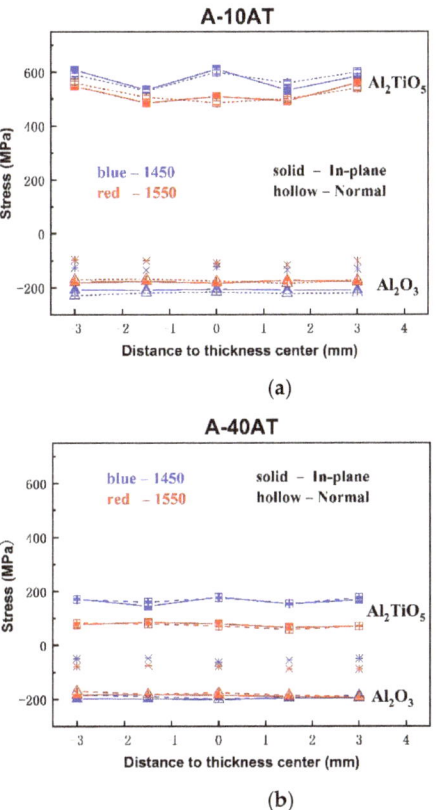

Figure 7. Mean phase stress profiles for Al$_2$O$_3$ (Δ) and Al$_2$TiO$_5$ (□) phases; through-thickness of Al$_2$O$_3$–Al$_2$TiO$_5$ bulk samples as a function of Al$_2$TiO$_5$ contents and sintering temperatures (1450 °C in blue line and 1550 °C in red line); values of $f_A\sigma_A + f_{AT}\sigma_{AT}$ are presented (✳) without a line. (**a**) A-10AT composites and (**b**) A-40AT composites.

As shown in Figure 7, the mean phase stress behaviors between the normal and in-plane directions are similar in all cases. The through-thickness stress profiles of both AT and Al$_2$O$_3$ phases are virtually flat, with residual tensile stresses in AT particulates and compressive stresses in the Al$_2$O$_3$ matrix. The high tensile stresses in AT were approximately 500–610 MPa for the A-10AT composites irrespective of the sintering temperature. It rapidly decreased as the AT content increased to approximately 80–180 MPa in the A-40AT composites. As the AT vol.% changed, the compressive stresses in the Al$_2$O$_3$ did not vary and remained at approximately −200 ± 30 MPa.

In addition, the residual stress behaviors according to different sintering treatments were studied. Regarding A–AT ceramics with the same AT addition, as the sintering temperature increased from 1450 to 1550 °C, tension in the AT phase decreased, and absolute compression values in the Al$_2$O$_3$ matrix slightly decreased in the A-10AT composites; however, no distinct variations were found in the A-40AT composites.

The residual thermal stresses in a particulate-reinforced composite are known to be caused by the elastic deformations of the matrix and particulates under uniform temperature change [45]. This is mainly due to the mismatch of thermal expansions and elastic constants between the matrix and particulates. For A–AT composites, the average crystallographic thermal expansion coefficient for the Al$_2$O$_3$ matrix is smaller than that for the AT inclusion [13,46,47]. Therefore, during fabrication, which is subjected to cooling from the maximum sintering temperature to room temperature, high compressive residual thermal stresses were induced in the Al_2O_3 matrix and tension in the AT inclusion.

Microstructural factors, such as particle volume fraction, size, shape, and microcracking, are known to also affect the magnitude and distribution of residual stresses. Considering the microstructure and grain size of AT and Al_2O_3 in A–AT samples with high AT contents, the grain size of AT was observed to increase. For a single AT particle, the surrounding misfit effects weakened because of the reduction in the contact area with the Al_2O_3 matrix and the increase in the contact area among AT particulates. The tension produced in AT due to the CTE misfit with the Al_2O_3 matrix correspondingly weakened. Furthermore, spontaneous AT particle microcracking occurred as the AT content increased to 40 vol.%. The occurrence of microcracking can relieve the stress energy in the AT phase. Both effects contributed to the remarkable decrease in tensile stresses in the AT phase of the A–AT composites with increasing AT content.

No significant change was detected in the stress value of the Al_2O_3 matrix as the addition of second-phase AT increased. This was inconsistent with the trend in other alumina-based ceramic composites reinforced by the second phase with higher thermal expansion, showing an increase in the absolute value of compression in the Al_2O_3 matrix [33,48]. This could be attributed to the special microcracking characteristics of AT particulates in A–AT ceramics. As AT vol.% increased, more microcracks were generated in AT and propagated along the boundaries of Al_2O_3 and AT. Microcracking benefited the absorption of stress energies during the expansion and shrinkage of materials during fabrication. Consequently, the stress state of the Al_2O_3 matrix in the A–AT composites was not significantly affected by the increase in AT content. This effect was also demonstrated by the thermal expansion curve of A–AT composites reported in a previous study [49], showing significant thermal expansion hysteresis in A-40AT compared with the A-10AT composites.

After sintering, temperature changes during cooling can lead to a higher CTE misfit strain value, thus increasing stresses. However, our measurement showed lower tension in the AT phase in the A–AT composites with higher sintering temperatures. Microcracking was presumed to be responsible for this inconsistency. Based on Figure 3 and Table 2, at higher sintering temperatures, the grain sizes of both AT and Al_2O_3 increase, resulting in further microcracking as a higher population of AT grains surpasses its critical size. This contributes to the release of thermal stress energy from samples with higher sintering temperatures; consequently, the tension in the AT phase decreases.

Moreover, according to the derived mean phase stresses of the Al_2O_3 matrix and AT particles derived by neutron diffraction, the macro-residual stresses, σ_{bulk}, of the A–AT bulk samples were calculated using the following equation:

$$\sigma_{bulk} = f_A \sigma_A + f_{AT} \sigma_{AT} \tag{11}$$

where f_A and f_{AT} are the volume fractions of Al_2O_3 and AT, respectively; and σ_A and σ_{AT} are the mean phase stresses of Al_2O_3 and AT, respectively.

The through-thickness values, σ_{bulk}, of the A–AT bulk samples are plotted in Figure 7. Macro-residual stresses, σ_{bulk}, were initially assumed negligible in all the A–AT bulk samples because no pressure was applied during the sintering process. However, according to the results, the σ_{bulk} values were not zero but compressive in the A–AT samples. This could be explained by considering the unknown stress states near the surface areas. As mentioned, strain scanning using neutron diffraction measurements was implemented 2 mm away from sample surfaces. The stress state near the surface was unknown because it could not be determined by neutron diffraction. Based on the hydrostatic assumption, tensile stress was anticipated around the surface area of the A–AT bulk.

3.4.2. Peak-Specific Residual Stresses

Peak-specific residual stresses were assessed by measuring the shifts in individual diffraction peaks in the TOF diffraction spectrum. Considering the peak intensity, as well as the non-overlapping and clear shape in the diffraction spectra, the peaks of Al_2O_3 (i.e., (030),

(116), (024), (113)) and AT (i.e., (243), (200), (043), and (113)) were selected for analysis, as shown in Figure 6. The d-spacing of each peak was obtained by Rietveld refinement. Peak-specific residual stresses for the selected peaks were calculated using Equations (9) and (10).

The through-thickness peak-specific residual stress behaviors of selected reflections of Al_2O_3 and AT in all samples, measured in both the in-plane and normal directions, are plotted in Figures 8 and 9, respectively.

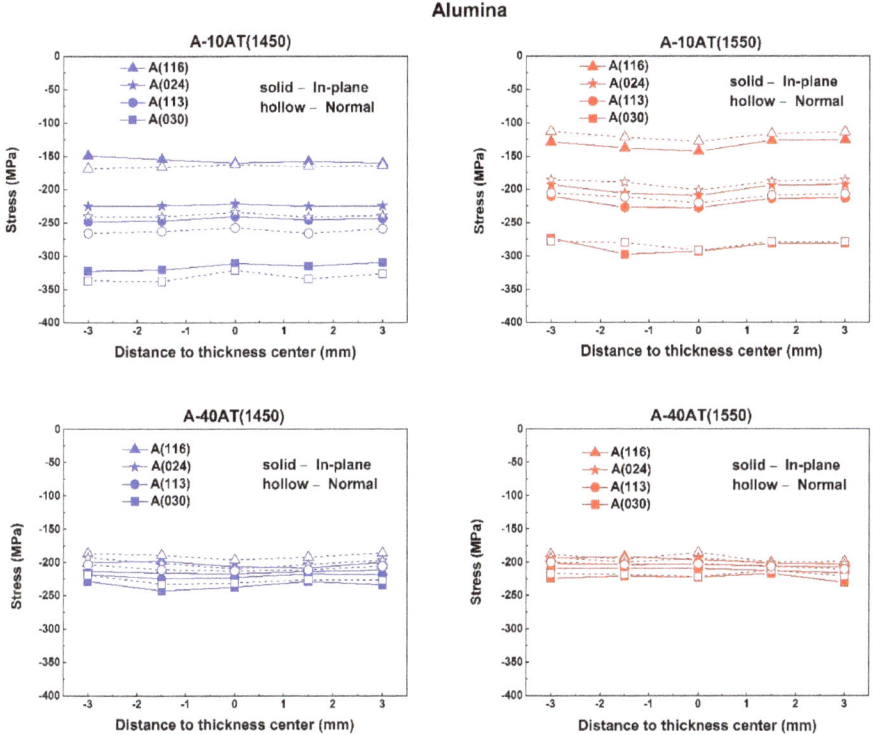

Figure 8. Through-thickness residual stress profiles corresponding to alumina reflections (i.e., (116), (024), (113), and (030)) in Al_2O_3–Al_2TiO_5 composites (1450 °C in blue line and 1550 °C in red line) measured along inplane (solid line and solid symbol) and normal (dash line and hollow symbol) directions.

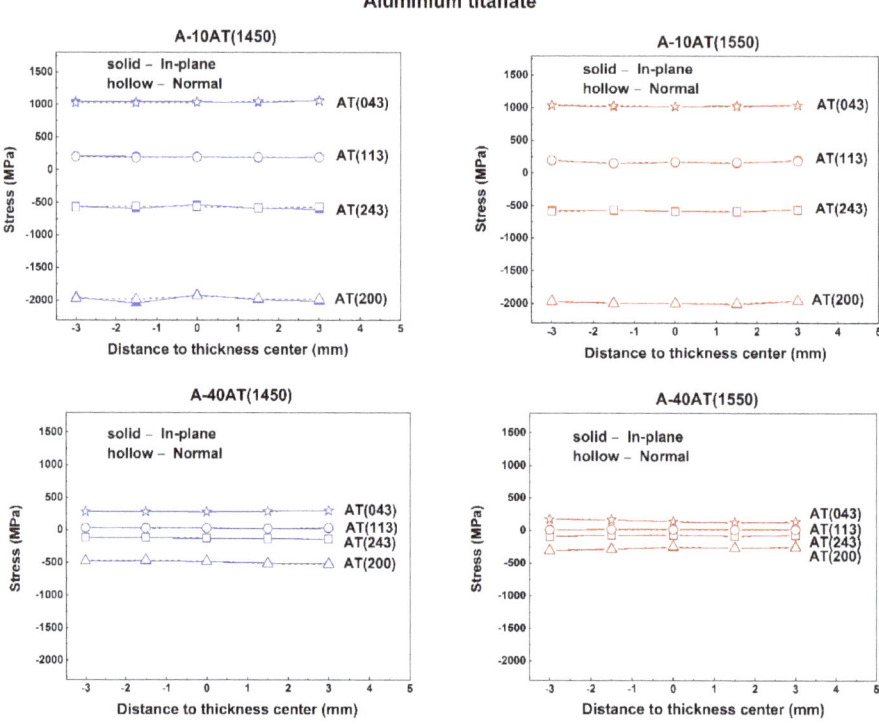

Figure 9. Through-thickness residual stress profiles corresponding to AT reflections (i.e., (043), (113), (243), and (200)) in Al_2O_3–Al_2TiO_5 composites (1450 °C in blue line and 1550 °C in red line) measured along in-plane (solid line and solid symbol) and normal (dash line and hollow symbol) directions.

As shown in Figures 8 and 9, the through-thickness residual stress profiles of all the selected reflections are virtually flat in both the in-plane and normal directions, confirming the occurrence of mean phase stress behaviors as previously discussed. Variations among peak-specific residual stresses are not remarkable between the in-plane and normal directions. The average peak-specific residual stress values of the selected peaks for both the Al_2O_3 and AT phases are summarized in Tables 4 and 5, respectively; these values were obtained regardless of the measurement directions. To clearly contrast the stresses of the Al_2O_3 and AT phases, their average mean phase stresses are also listed.

Table 4. Average residual stress values of selected peaks for Al_2O_3 compared with corresponding mean phase stresses.

Sample	Mean Phase Stress σ_{Al2O3} (MPa)	Peak-Specific Residual Stress (MPa)			
		Al_2O_3 (030)	Al_2O_3 (116)	Al_2O_3 (024)	Al_2O_3 (113)
A-10AT (1450)	−208 ± 7	−324 ± 12	−160 ± 6	−232 ± 8	−253 ± 9
A-10AT (1550)	−176 ± 6	−283 ± 7	−125 ± 9	−195 ± 8	−215 ± 7
A-40AT (1450)	−194 ± 6	−230 ± 6	−196 ± 7	−207 ± 8	−14 ± 7
A-40AT (1550)	−183 ± 7	−219 ± 4	−195 ± 5	−201 ± 4	−208 ± 5

Table 5. Average residual stress values of selected peaks for AT phase compared with corresponding mean phase stresses.

Sample	Mean Phase Stress σ_{AT} (MPa)	Peak-Specific Residual Stress (MPa)			
		AT (243)	AT (200)	AT (043)	AT (113)
A-10AT (1450)	576 ± 30	−572 ± 20	−1977 ± 34	1041 ± 10	195 ± 8
A-10AT (1550)	518 ± 29	−579 ± 10	−1985 ± 18	1027 ± 9	171 ± 16
A-40AT (1450)	166 ± 11	−128 ± 8	−493 ± 21	282 ± 7	27 ± 4
A-40AT (1550)	74 ± 8	−79 ± 7	−274 ± 18	145 ± 15	14 ± 2

For the same phase in the same sample, the obtained peak-specific residual stress values varied among different *hkl* reflections. Compression was developed in all the selected *hkl* reflections of the Al_2O_3 phase in the studied A–AT composites. However, the AT phase underwent tension in the reflections of (043) and (113); compression occurred in (243) and (200). The values of the peak-specific residual stresses in each phase differed from their mean phase stresses. As the AT content increased from 10 to 40 vol.%, the variation ranges of residual stresses in different reflections became smaller for both the Al_2O_3 and AT phases. However, the orders of peak-specific residual stress values in different *hkl* reflections were unchanged in each phase.

The effects of different sintering temperatures on peak-specific residual stresses were investigated. With a higher sintering temperature, the A-10AT composites exhibited reduced absolute values of compression in all selected *hkl* reflections of Al_2O_3; however, no distinct difference in peak-specific residual stresses of the AT phase was observed. Conversely, in the A-40AT composites, the peak-specific residual stresses in the Al_2O_3 phase between A-40AT(1450) and A-40AT(1550) were similar, whereas the AT phase exhibited evident differences in peak-specific residual stresses between these two samples. With a higher sintering temperature in the A-40AT composites, the variation range of residual stresses in different *hkl* reflections were reduced in the AT phases.

The results of peak-specific residual stresses, calculated from the d-spacing of each reflection, can be explained by the lattice parameter values. For hexagonal α-Al_2O_3, the relationship between the d-spacing (d_{hkl}) for a given *hkl* plane and the lattice parameters, *a* ($a = b$) and *c*, is written as follows [50]:

$$d_{hkl} = \frac{1}{\sqrt{\frac{4}{3a^2}(h^2 + k^2 + hk) + \frac{l^2}{c^2}}} \quad (12)$$

According to the obtained lattice parameters listed in Table 3, the *a* and *c* axes of α-Al_2O_3 shrunk in all the studied A–AT composites when compared with those of the Al_2O_3 reference powder. Thus, according to Equation (12), for a given lattice plane *hkl* of Al_2O_3, the d-spacing decreased, and compression developed in all the selected reflections in the Al_2O_3 phase in the A–AT composites.

As the AT content increased from 10 to 40 vol.%, the *a* axis of Al_2O_3 slightly expanded, but the *c* axis shrunk. This led to peak-specific residual stresses in some *hkl* reflections and a certain reduction in the Al_2O_3 phase. For example, for *hk0* planes in the Al_2O_3 phase in the A-40AT composites, with a slightly increased *a* value, d_{hk0} increased, and the absolute values of the compressive peak-specific residual stresses decreased compared with those in the A-10AT composites. This is consistent with the measured residual stresses of A(030) in the A–AT composites, as listed in Table 4.

Compared with the A-10AT(1450) composites, the A-10AT(1550) composites showed slight increases in the *a* and *c* axes of α-Al_2O_3. Thus, with a higher sintering temperature value, the d-spacing of each Al_2O_3 reflection slightly increased, and the absolute values of compressive peak-specific residual stresses decreased. In the A-40AT composites, the difference in the lattice parameters of Al_2O_3 was not distinct at different sintering temperatures; thus, no remarkable difference was observed in the peak-specific residual stresses of Al_2O_3 in the two samples.

For orthorhombic AT, with the lattice parameters a, b, and c, the interplanar spacing (d_{hkl}) is given by the following:

$$d_{hkl} = \frac{1}{\sqrt{\frac{h^2}{a^2} + \frac{k^2}{b^2} + \frac{l^2}{c^2}}} \quad (13)$$

As summarized in Table 3, the a axis of the AT phase shrunk, and the b and c axes expanded in all the studied A–AT composites compared with those of the AT reference powder. Thus, for a specific hkl plane of AT, the d-spacing may decrease or increase according to Equation (13). This leads to peak-specific residual stresses, either compression or tension in different selected reflections in AT. For example, for the $0kl$ reflections, d_{0kl} increased, and tension developed. For the $h00$ reflections of AT, d_{h00} decreased, and compression developed.

As the AT content increased from 10 to 40 vol.%, the a axis of the AT phase expanded, and the b and c axes shrunk. Thus, for the $0kl$ reflections in AT, the d-spacing decreased, and tension decreased. For the $h00$ reflections in AT, d_{h00} increased, and the absolute values of compression decreased. These are all in accordance with the measured peak-specific residual stress behaviors of AT(043) and (200), as summarized in Table 5.

With regard to the effect of different sintering temperatures, owing to the similar lattice parameters of AT in the A-10AT(1450) and A-10AT(1550) samples, the peak-specific residual stresses for each selected AT reflection were similar in both samples. In the A-40AT composites, as the sintering temperature increased from 1450 to 1550 °C, the a axis of the AT phase expanded, and the b and c axes shrunk. Consequently, for AT(430) in A-40AT(1550), d_{043} decreased, and the tension decreased with respect to that in A-40AT(1450). For AT(200) in A-40AT(1550), d_{200} increased; consequently, the absolute compression value decreased with respect to that in A-40AT(1450).

Owing to the strong anisotropy of thermal expansion in the AT phase, with higher AT content and sintering temperature, more microcracks are generated in the A–AT composites. This leads to the reduced connectivity of the material in the composites; thus, the restraints in the material are weakened. Hence, the anisotropy of peak-specific residual stresses in both Al_2O_3 and AT phases was weaker in A-40AT than in the A-10AT composites, and the anisotropy of the peak-specific residual stresses of the AT phase was weaker in A-40AT(1550) than in A-40AT(1450).

As mentioned above, it is well accepted that peak-specific residual stress behaviors are mainly determined by the crystal structure of each phase; this is closely related to the CTE anisotropy in each phase and the microstructure of materials. On the one hand, owing to the anisotropy of thermal expansions and elastic properties in each hkl direction, for the same sample and phase, the residual stress values obtained vary from different reflections. The sign and magnitude of residual stress values considerably depend on the reflection used for analysis in the diffraction method. This is an important concern in residual stress measurements and analysis using the diffraction method with single-peak reflections in complex materials. On the other hand, with looser microstructures, such as microcracking, the anisotropy of peak-specific residual stresses in each phase was distinctly weakened. Deriving reliable properties is beneficial owing to lower internal stresses. However, a looser structure may limit the strength of materials. Thus, to balance the weakening of the anisotropy of residual stresses and improve mechanical properties, optimizing and controlling the microstructure of materials are crucial.

4. Summary and Conclusions

In summary, a series of Al_2O_3–Al_2TiO_5 bulk composites was investigated, with different AT contents (10 and 40 vol.%) and fabricated with different sintering temperature (1450 and 1550 °C) in order to develop significantly different microstructural features. The microstructure, crystal structure, and through-thickness residual stress profiles of the samples were obtained.

The main conclusions of the present study are as follows:

(1) The sintering temperatures of 1450 °C ensured the complete formation and retention of Al_2TiO_5 during the reaction sintering and the subsequent cooling of the A–AT composites. Some decomposition of AT occurred in the A–AT composites at the sintering temperature of 1550 °C.

(2) The increase of AT content (from 10 to 40 vol.%) and sintering temperature (from 1450 to 1550 °C) resulted in microstructure evolution in the A–AT composites, from a dense fine-grained microstructure to coarse microstructure with AT grain growth and microcracking. Therefore, the flexure strength was correspondingly decreased.

(3) The lattice parameters of both Al_2O_3 and AT varied with the AT content and sintering temperature in the studied A–AT composites, mainly because of thermal strains.

(4) Owing to the CTE mismatch between the matrix and particles, tensile residual stresses developed in the AT particulates and compressive stresses in the Al_2O_3 matrix, with almost flat through-thickness residual stress profiles. The increase of AT content and sintering temperature led to the change of the mean phase stress of both AT and Al_2O_3 in the composites. Microcracking is an important factor of residual stress, contributing to the release of thermal stress energy in samples.

(5) Owing to the thermal expansion anisotropy in each phase, significant stresses built up in different hkl reflections for both the Al_2O_3 and AT phases. The peak-specific residual stress profiles of both the Al_2O_3 matrix and AT particulates were virtually flat throughout the sample thickness, with stress values varying along different hkl reflections. The sign and magnitude of residual stress values considerably depend on the reflection used for analysis in the diffraction method. This is an important concern in residual stress measurements and analysis using the diffraction method with single-peak reflections in complex materials.

Author Contributions: Conceptualization, K.F., J.R.-H. and C.B.; methodology, J.R.-H., K.F.; software, W.J.; validation, W.J., S.B. and H.Z.; formal analysis, K.F. and J.R.-H.; investigation, K.F., J.R.-H. and W.J.; resources, J.R.-H., C.B. and S.B.; data curation, K.F.; writing—original draft preparation, K.F. and W.J.; writing—review and editing, K.F. and S.B.; visualization, H.Z.; supervision, P.Y.; project administration, W.F.; funding acquisition, K.F. and W.F. All authors have read and agreed to the published version of the manuscript.

Funding: This research was funded by the National Natural Science Foundation of China, grant number 51805445, 52175209; the Chunhui Plan Cooperative Scientific Research Project of the Ministry of Education of China, grant number 191659; the Opening Foundation of Sichuan Engineering Research Center for Powder Metallurgy, grant number SC-FMYJ2019-04. Beamtime was granted by ISIS ENGIN-X, Exp. No. RB1920330.

Data Availability Statement: Some or all data generated or used during the study are available from the corresponding author by request.

Acknowledgments: All financial support is greatly appreciated. The ISIS neutron and muon source are thanked for the provision of beamtime for residual strain scanning.

Conflicts of Interest: The authors declare no conflict of interest.

References

1. Borrell, A.; Salvador, M.D.; Rocha, V.G.; Fernández, A.; Molina, T.; Moreno, R. Enhanced properties of alumina–aluminium titanate composites obtained by spark plasma reaction-sintering of slip cast green bodies. *Compos. Part B Eng.* **2013**, *47*, 255–259. [CrossRef]
2. Maki, R.S.; Suzuki, Y. Mechanical strength and electrical conductivity of reactively-sintered pseudobrookite-type Al_2TiO_5–$MgTi_2O_5$ solid solutions. *J. Ceram. Soc. Jpn.* **2016**, *124*, 1–6. [CrossRef]
3. Kim, I.J. Thermal stability of Al_2TiO_5 ceramics for new diesel particulate filter applications-a literature review. *J. Ceram. Process. Res.* **2010**, *11*, 411–418.
4. Papitha, R.; Suresh, M.B.; Johnson, R.; Dibakar, D. High-Temperature Flexural Strength and Thermal Stability of Near Zero Expanding doped Aluminum Titanate Ceramics for Diesel Particulate Filters Applications. *Int. J. Appl. Ceram. Technol.* **2013**, *11*, 773–782. [CrossRef]

5. Cong, C.A.; Brm, A.; Cp, A.; Js, A.; If, A.; Gba, B. The correlation between porosity characteristics and the crystallographic texture in extruded stabilized aluminium titanate for diesel particulate filter applications. *J. Eur. Ceram. Soc.* **2020**, *40*, 1592–1601.
6. Alves, P.; Rodrigues, M.F.; Cossu, C.M.F.A.; Magnago, R.d.O.; Ramos, A.S.; dos Santos, C. Characterization of Al_2O_3-Al_2TiO_2 Ceramic Composites: Effects of Sintering Parameters on the Properties. In *Materials Science Forum*; Trans Tech Publications Ltd.: Freienbach, Switzerland, 2018; pp. 118–123.
7. Skala, R.D.; Manurung, P.; Low, I.M. Microstructural design, characterisation and indentation responses of layer-graded alumina/aluminium–titanate composites. *Compos. Part B Eng.* **2006**, *37*, 466–480. [CrossRef]
8. Hu, Y.; Li, M.Z.; Shen, Q.; Xu, W.P. Thermal Shock-Resistance Performance of Al_2TiO_5/Al_2O_3 Composites. In *Advanced Materials Research*; Trans Tech Publications Ltd.: Freienbach, Switzerland, 2009; pp. 108–111.
9. Botero, C.A.; Jiménez-Piqué, E.; Baudín, C.; Salán, N.; Llanes, L. Nanoindentation of Al_2O_3/Al_2TiO_5 composites: Small-scale mechanical properties of Al_2TiO_5 as reinforcement phase. *J. Eur. Ceram. Soc.* **2012**, *32*, 3723–3731. [CrossRef]
10. de Portu, G.; Bueno, S.; Micele, L.; Baudin, C.; Pezzotti, G. Piezo-spectroscopic characterization of alumina-aluminum titanate laminates. *J. Eur. Ceram. Soc.* **2006**, *26*, 2699–2705. [CrossRef]
11. Padture, N.P.; Bennison, S.J.; Chan, H.M. Flaw-Tolerance and Crack-Resistance Properties of Alumina-Aluminum Titanate Composites with Tailored Microstructures. *J. Am. Ceram. Soc.* **1993**, *76*, 2312–2320. [CrossRef]
12. Asmi, D.; Low, I.M. Processing of an in-situ Layered and Graded Alumina/Calcium-Hexaluminate composite: Physical Characteristics. *J. Eur. Ceram. Soc.* **1998**, *18*, 2019–2024. [CrossRef]
13. Skala, R.D.; Li, D.; Low, I.M. Diffraction, structure and phase stability studies on aluminium titanate. *J. Eur. Ceram. Soc.* **2009**, *29*, 67–75. [CrossRef]
14. Tsetsekou, A. A comparison study of tialite ceramics doped with various oxide materials and tialite–mullite composites: Microstructural, thermal and mechanical properties. *J. Eur. Ceram. Soc.* **2005**, *25*, 335–348. [CrossRef]
15. Chan, H.M. Layered Ceramics: Processing and Mechanical Behavior. *Annu. Rev. Mater. Res.* **2003**, *27*, 249–282. [CrossRef]
16. Naglieri, V.; Palmero, P.; Montanaro, L.; Chevalier, J. Elaboration of alumina-zirconia composites: Role of the zirconia content on the microstructural and mechanical properties. *Materials* **2013**, *6*, 2090–2102. [CrossRef]
17. Magnani, G.; Brillante, A. Effect of the composition and sintering process on mechanical properties and residual stresses in zirconia-alumina composites. *J. Eur. Ceram. Soc.* **2005**, *25*, 3383–3392. [CrossRef]
18. Park, S.Y.; Jung, S.W.; Chung, Y.B. The effect of starting powder on the microstructure development of alumina–aluminum titanate composites. *Ceram. Int.* **2003**, *29*, 707–712. [CrossRef]
19. Chen, C.-H.; Awaji, H. Temperature dependence of mechanical properties of aluminum titanate ceramics. *J. Eur. Ceram. Soc.* **2007**, *27*, 13–18. [CrossRef]
20. Singh, M.; Manurung, P.; Low, I.M. Depth profiling of near-surface information in a functionally graded alumina/aluminium titanate composite using grazing-incidence synchrotron radiation diffraction. *Mater. Lett.* **2002**, *55*, 344–349. [CrossRef]
21. Bueno, S.; Baudín, C. Design and processing of a ceramic laminate with high toughness and strong interfaces. *Compos. Part A Appl. Sci. Manuf.* **2009**, *40*, 137–143. [CrossRef]
22. Fan, K.; Ruiz-Hervias, J.; Pastor, J.Y.; Gurauskis, J.; Baudín, C. Residual stress and diffraction line-broadening analysis of Al_2O_3/Y-TZP ceramic composites by neutron diffraction measurement. *Int. J. Refract. Met. Hard Mater.* **2017**, *64*, 122–134. [CrossRef]
23. Withers, P.J. Mapping residual and internal stress in materials by neutron diffraction. *Comptes Rendus Phys.* **2007**, *8*, 806–820. [CrossRef]
24. Bueno, S.; Moreno, R.; Baudin, C. Reaction sintered Al_2O_3/Al_2TiO_5 microcrack-free composites obtained by colloidal filtration. *J. Eur. Ceram. Soc.* **2004**, *24*, 2785–2791. [CrossRef]
25. Santisteban, J.; Daymond, M.; James, J.; Edwards, L. ENGIN-X: A third-generation neutron strain scanner. *J. Appl. Crystallogr.* **2006**, *39*, 812–825. [CrossRef]
26. Edwards, L. 14 Near-surface stress measurement using neutron diffraction. In *Analysis of Residual Stress by Diffraction Using Neutron and Synchrotron Radiation*; CRC Press: Boca Raton, FL, USA, 2003; p. 233.
27. Coelho, A. TOPAS-Academic V5. Coelho Software, Brisbane, Australia. 2012. Available online: https://www.topas-academic.net/ (accessed on 7 October 2021).
28. Daymond, M.; Bourke, M.; Von Dreele, R.; Clausen, B.; Lorentzen, T. Use of Rietveld refinement for elastic macrostrain determination and for evaluation of plastic strain history from diffraction spectra. *J. Appl. Phys.* **1997**, *82*, 1554–1562. [CrossRef]
29. Balzar, D.; Popa, N.C. Analyzing microstructure by Rietveld refinement. *Rigaku J.* **2005**, *22*, 16.
30. McCusker, L.; Von Dreele, R.; Cox, D.; Louer, D.; Scardi, P. Rietveld refinement guidelines. *J. Appl. Crystallogr.* **1999**, *32*, 36–50. [CrossRef]
31. Lewis, J.; Schwarzenbach, D.; Flack, H. Electric field gradients and charge density in corundum, α-Al_2O_3. *Acta Crystallogr. Sect. A Cryst. Phys. Diffr. Theor. Gen. Crystallogr.* **1982**, *38*, 733–739. [CrossRef]
32. Choo, H.; Bourke, M.; Nash, P.; Daymond, M.; Shi, N. Thermal residual stresses in NiAl–AlN–Al_2O_3 composites measured by neutron diffraction. *Mater. Sci. Eng. A* **1999**, *264*, 108–121. [CrossRef]
33. Fan, K.; Ruiz-Hervias, J.; Gurauskis, J.; Sanchez-Herencia, A.J.; Baudín, C. Neutron diffraction residual stress analysis of Al_2O_3/Y-TZP ceramic composites. *Boletín Soc. Española Cerámica Vidr.* **2016**, *55*, 13–23. [CrossRef]

34. Oliver, E.C. The Generation of Internal Stresses in Single and Two Phase Materials. Ph.D. Thesis, University of Manchester, Manchester, UK, 2002.
35. Gnäupel-Herold, T. ISODEC: Software for calculating diffraction elastic constants. *J. Appl. Crystallogr.* **2012**, *45*, 573–574. [CrossRef]
36. Gnäupel-Herold, T. A software for diffraction stress factor calculations for textured materials. *Powder Diffr.* **2012**, *27*, 114–116. [CrossRef]
37. Manurung, P.; Low, I.M.; O'Connor, B.H.; Kennedy, S. Effect of β-spodumene on the phase development in an alumina/aluminium-titanate system. *Mater. Res. Bull.* **2005**, *40*, 2047–2055. [CrossRef]
38. Violini, M.A.; Hernández, M.; Conconi, M.S.; Suárez, G.; Rendtorff, N.M. A dynamic analysis of the aluminum titanate (Al_2TiO_5) reaction-sintering from alumina and titania, properties and effect of alumina particle size. *J. Therm. Anal. Calorim.* **2021**, *143*, 95–101. [CrossRef]
39. Keyvani, N.; Azarniya, A.; Hosseini, H.R.M.; Abedi, M.; Moskovskikh, D. Thermal stability and strain sensitivity of nanostructured aluminum titanate (Al2TiO5). *Mater. Chem. Phys.* **2019**, *223*, 202–208. [CrossRef]
40. Low, I.M.; Lawrence, D.; Jones, A.; Smith, R.I. Dynamic Analyses of the Thermal Stability of Aluminium Titanate by Time-Of-Flight Neutron Diffraction. In *Developments in Advanced Ceramics and Composites: Ceramic Engineering and Science Proceedings*; John Wiley & Sons: Hoboken, NJ, USA, 2008; Volume 26, p. 8.
41. Low, I.M.; Oo, Z. Effect of Grain Size and Controlled Atmospheres on the Thermal Stability of Aluminium Titanate. In *AIP Conference Proceedings*; American Institute of Physics: College Park, MD, USA, 2010.
42. Fan, K.; Pastor, J.Y.; Ruiz-Hervias, J.; Gurauskis, J.; Baudin, C. Determination of mechanical properties of Al_2O_3/Y-TZP ceramic composites: Influence of testing method and residual stresses. *Ceram. Int.* **2016**, *42*, 18700–18710. [CrossRef]
43. Uribe, R.; Baudín, C. Influence of a dispersion of aluminum titanate particles of controlled size on the thermal shock resistance of alumina. *J. Am. Ceram. Soc.* **2003**, *86*, 846–850. [CrossRef]
44. Low, I.M.; Oo, Z. Dynamic neutron diffraction study of thermal stability and self-recovery in aluminium titanate. In *Strategic Materials and Computational Design: Ceramic Engineering and Science Proceedings*; American Ceramic Society: Columbus, OH, USA, 2010; Volume 31.
45. Taya, M.; Hayashi, S.; Kobayashi, A.S.; Yoon, H. Toughening of a Particulate-Reinforced Ceramic-Matrix Composite by Thermal Residual Stress. *J. Am. Ceram. Soc.* **1990**, *73*, 1382–1391. [CrossRef]
46. Bartolome, J.F.; Requena, J.; Moya, J.S.; Li, M.; Guiu, F. Cyclic fatigue crack growth resistance of Al_2O_3-Al_2TiO_5 composites. *Acta Mater.* **1996**, *44*, 1361–1370. [CrossRef]
47. de Arenas, I.B. Reactive sintering of aluminum titanate. In *Sintering of Ceramics-New Emerging Techniques*; InTech: London, UK, 2012.
48. Mori, T.; Tanaka, K. Average stress in matrix and average elastic energy of materials with misfitting inclusions. *Acta Metall.* **1973**, *21*, 571–574. [CrossRef]
49. Hernández, M.; Bueno, S.; Sánchez, T.; Anaya, J.J.; Baudín, C. Non-destructive characterisation of alumina/aluminium titanate composites using a micromechanical model and ultrasonic determinations: Part II. Evaluation of microcracking. *Ceram. Int.* **2008**, *34*, 189–195. [CrossRef]
50. Kelly, A.A.; Knowles, K.M. *Crystallography and Crystal Defects*; John Wiley & Sons: Hoboken, NJ, USA, 2012.

Article

Studies on Electron Escape Condition in Semiconductor Nanomaterials via Photodeposition Reaction

Chen Ye [1] and Yu Huan [2,*]

[1] School of Chemistry and Chemical Engineering, University of Jinan, No. 336, West Road of Nan Xinzhuang, Jinan 250022, China; yechen-chem@hotmail.com

[2] School of Material Science and Engineering, University of Jinan, No. 336, West Road of Nan Xinzhuang, Jinan 250022, China

* Correspondence: mse_huany@ujn.edu.cn

Abstract: In semiconductor material-driven photocatalysis systems, the generation and migration of charge carriers are core research contents. Among these, the separation of electron-hole pairs and the transfer of electrons to a material's surface played a crucial role. In this work, photodeposition, a photocatalysis reaction, was used as a "tool" to point out the electron escaping sites on a material's surface. This "tool" could be used to visually indicate the active particles in photocatalyst materials. Photoproduced electrons need to be transferred to the surface, and they will only participate in reactions at the surface. By reacting with escaped electrons, metal ions could be reduced to nanoparticles immediately and deposited at electron come-out sites. Based on this, the electron escaping conditions of photocatalyst materials have been investigated and surveyed through the photodeposition of platinum. Our results indicate that, first, in monodispersed nanocrystal materials, platinum nanoparticles deposited randomly on a particle's surface. This can be attributed to the abundant surface defects, which provide driving forces for electron escaping. Second, platinum nanoparticles were found to be deposited, preferentially, on one side in heterostructured nanocrystals. This is considered to be a combination result of work function difference and existence of heterojunction structure.

Keywords: photodeposition; electron escape; semiconductor photocatalyst

---

1. Introduction

Since the breakthrough development of photocatalysis was reported in 1972 [1], semiconductor material has been studied widely and deeply. These studies not only focused on the invention of new composition, new construction, and functionalized materials but also on the in-depth analysis of their photocatalytic reaction mechanism [2–5]. The overall semiconductor-driven photocatalytic process includes three steps: (1) the generation of electron-hole charge carriers under the irradiation of a light source; (2) the separation of electrons and holes; (3) the migration of electrons to the reactive sites on the crystal's surface [6]. Many explanations have been proposed to elucidate the mechanism of the separation of electrons and holes [7]. However, it is difficult to afford direct evidence of the electron motion, despite that an efficient electron escape module could promote the charge carrier separation. Moreover, the photo-induced catalytic reactions will only occur when the electrons come out of the surface. Thus, the studies on how and where the electrons escape from the materials make more sense.

In monodispersed semiconductor nanocrystals, surface engineering is one of the most efficient strategies to overcome the limitations of semiconductor materials. It is practical to improve the properties, as well as application performance of semiconductor nanomaterials, via surface modification and functionalization. Surface engineering of semiconductor materials includes the studies on geometry effects, surface defects, capping ligands, decomposition of nanoparticles or the nanoshell, and so on [8–16]. During the last decades, studies on geometry effects on nanocrystal materials have attracted strong interest as a result of the

possibility to tailor the materials. Studies on geometry effects of nanomaterials revealed the effects of size, morphology, and surface structure for nanoparticles. Surface geometry effects directly affect the surface properties, including work function and surface energy, which will further affect the particle properties and surface functionalization [13–16]. Meanwhile, surface defect engineering has also been considered as a useful approach for the modification of electronic and chemical properties of semiconductor nanomaterials, which enhances their activity photocatalysts [17–20]. Until now, photocatalyst materials, with various types of and abundant defects, have been studied, including metal oxides, metal chalcogenides, graphene materials, etc. [21–26]. Some results demonstrated that surface defects may serve as electron traps, making the electrons migrate to a more reactive site, or directly out of the surface [27–30]. Subsequently, reactions will occur between the surface adsorbed reactants and the escaped electrons.

Different from monodisperse nanoparticles, heterostructured nanomaterials, which can be formed by loading metal particles or metallic compounds on a surface, are completely other mechanisms. The heterojunction can be a p-n junction or a Schottky barrier. Free electrons existing in these materials were driven to migrate across the junction [31–33]. Therefore, the electrons should prefer to escape from the surface of electron attractors such as p-type materials or metal particles.

In an electrolytic reaction system, only the solid and gas phase products can be easily detected since the solid products will be deposited on the electrode, while gas products can be collected in special vessels. Similar to this, if we intend to observe the electron escape sites on the material surface, the formation of solid products could be a great option. Photodeposition, a method based on the photocatalytic property of semiconductors, is usually used to prepared metal-loading semiconductor materials [34–36]. The metal ions, which adsorbed on the surface of semiconductor materials, will react with electrons once they come out from surface. Theoretically, only the metal ions, which adsorbed at or near the electron escaping sites, can be reduced. Furthermore, metal particles will only be deposited at electron escaping sites if the metal ions are small enough.

In this work, we have studied the electron escaping position via photodeposition of platinum nanoparticles. The electron escaping conditions were identified in different kinds of photocatalyst materials, including the different types of monodisperse semiconductor nanocrystals and the crystals with a special heterojunction structure. The results revealed that surface defects did have good electron trap ability. In addition, the existence of heterojunction also plays an important role for electron motion.

2. Materials & Methods

Chemicals: titanium oxide mix phase nanofiber (TiO_2 MP, 99%), titanium oxide nanofiber type 1 (TiO_2 nanofiber type 1, 99%), titanium oxide nanofiber type 2 (TiO_2 nanofiber type 2, 99%), copper sulfide nanosheets (CuS), star-like $BiVO_4$, flower-like Bi_2WO_6, and platinum (IV) chloride ($PtCl_4$, 98%) were purchased from commercial sources and used without further purification.

Photodeposition of Pt: the photodeposition of Pt nanoparticles was taken under ultraviolet (UV) light illumination. The photocatalyst materials were dispersed in deionized water and mixed with $PtCl_4$ aqueous solution. The mixture, with continuous magnetic stirring, was irradiated with ultraviolet light (λ = 365 nm) from a UV lamp. The light intensity on the sample was 134 mW/cm^2. For comparison, the reactions without irradiation were also prepared.

Characterization: transmission electron microscopes (TEM), high-resolution transmission electron microscopy (HRTEM), and selected area electron diffraction (SAED) were carried out on a JEOL 2100F electron microscope (JEOL, Tokyo, Japan), operated with an accelerating voltage of 200 kV. Energy-dispersive X-ray (EDX) analysis and corresponding elemental mapping data were taken with the X-ray spectroscopy (Oxford X-Max 80, Oxford Instruments, Abingdon, UK) attached to the TEM instrument. To prepare the TEM specimens, a drop of nanocrystals, dispersed in ethanol, was dropped on the surface of a lacey formvar/carbon 200-mesh Cu grid. X-ray diffraction analysis (XRD) patterns were collected on a Rigaku SmartlabSE X-ray diffractometer (Rigaku, Tokyo, Japan) by using Cu Kα radiation.

3. Results

In this work, the photodeposition of Pt nanoparticles were employed on different kinds of photocatalyst materials, including TiO_2, Bi_2WO_6, $BiVO_4$, and CuS. The reaction times and the amount of $PtCl_4$ were summarized in Table 1. All the photocatalyst materials in this work were prepared by hydrothermal method. The TiO_2 nanomaterials used in this work included three types: pure phase TiO_2 nanofiber type 1 (TiO_2 t1), pure phase TiO_2 nanofiber type 2 (TiO_2 t2), and mixed phase TiO_2 (TiO_2 MP). The TiO_2 MP were synthesized in a one-pot reaction. The crystal structure of the purchased TiO_2 were characterized by the XRD patterns in Figure 1. It can be observed that the XRD peaks of TiO_2 t1 coincided well with the brookite phase (JCPDS No. 46-1238), and the XRD peaks of TiO_2 t2 were indexed to the anatase phase (JCPDS No. 21-1272) without any secondary phase. The broadened XRD peaks should be correlated with the nano-scale particles. The XRD peaks of the TiO_2 MP sample at 14.2°, 24.9°, 28.6°, 43.5°, and 44.5° were ascribed to the brookite phase, while the peaks at 25.3°, 37.8°, and 48.0° were attributed to the anatase phase. It confirmed that the brookite and anatase phases coexisted in the TiO_2 MP sample.

Table 1. Summary of the crystal structure of Pt deposited nanocrystals and the reaction conditions.

Basic Materials	Crystal Structure	Pt^{4+} Loading Amount	Reaction Time
TiO_2 t1	Brookite	1.2–12.0 wt%	30 min
TiO_2 t2	Anatase	1.2–12.0 wt%	30 min
TiO_2 MP	Brookite + Anatase	0.23–12.0 wt%	15 min–2 h
Bi_2WO_6	Russellite	2.9 wt%	30 min
Bi_2VO_4	Clinobisvanite	2.9 wt%	30 min
CuS	Covellite	2.9 wt%	30 min

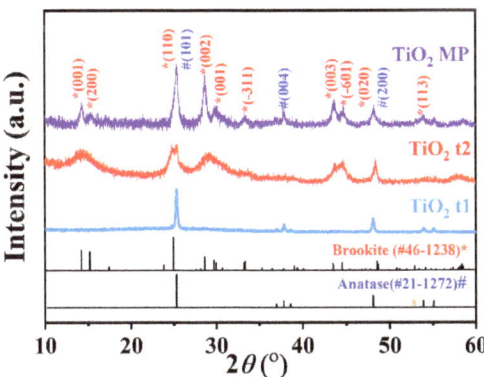

Figure 1. XRD patterns of TiO_2 MP, TiO_2 t1, and TiO_2 t2 samples. The peaks marked by pound sign and asterisk are dependent to anatase phase (JCPDS No. 21-1272) and brookite phase (JCPDS No. 46-1238), respectively.

Figure 2 showed the XRD results of Pt-deposited Bi_2WO_6, $BiVO_4$, and CuS photocatalyst materials. The crystal structure of Bi_2WO_6, $BiVO_4$, and CuS nanomaterials could be illustrated by XRD patterns in Figure 2. The XRD patterns of Bi_2WO_6, $BiVO_4$, and CuS were well indexed to the russellite phase (JCPDS No. 39-0256), clinobisvanite phase (JCPDS No. 14-0688), and covellite phase (JCPDS No. 06-0464), respectively. These XRD results proved that these nanomaterials had a good crystallization crystal structure and the expected stoichiometric composition. The broadened XRD peaks should be correlated with the nano-scale particles. Since the experiments were attended to load Pt nanoparticles on these photocatalyst materials, typical XRD peaks, located at 39.8° and 46.2° of platinum (JCPDS No. 04-0802), were also marked in the Figure 2. Disappointingly, none of the Pt XRD peaks were detected in these four Pt-deposited photocatalyst materials. This might be because of the small size and low content of Pt nanoparticles.

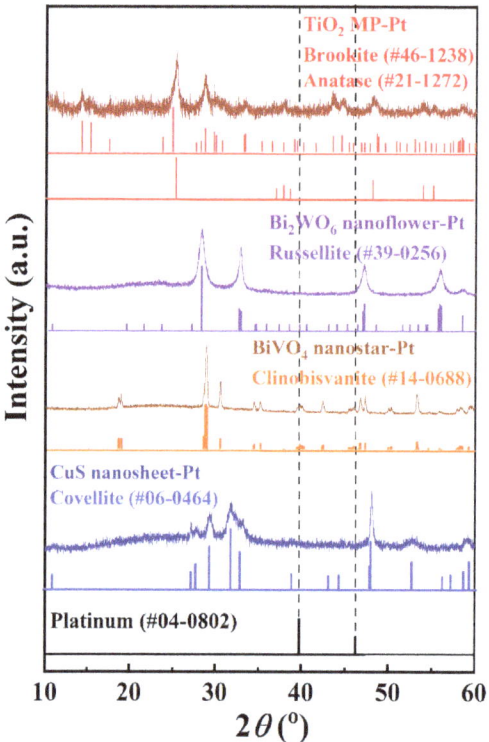

Figure 2. XRD patterns of Pt-deposited TiO_2 MP nanofiber, Pt-deposited Bi_2WO_6 nanoflower, Pt-deposited $BiVO_4$ nanostar, and Pt-deposited CuS nanosheet.

In order to confirm the existing of Pt nanoparticles, TEM, HRTEM, SAED, and EDX analyses were carried out. Figure 3 showed the TEM images of Pt-deposited TiO_2 MP (a), Bi_2WO_6 (b), $BiVO_4$ (c), CuS (d), TiO_2 t1 (e), and TiO_2 t2 (f). The black dots on the particles, with the size of 1–2 nm, were considered as the Pt nanoparticles. The Pt nanoparticles were homogeneously distributed on the surface of the photocatalyst matrix. In addition, the deposited Pt nanoparticles increased with the increasing $PtCl_4$ solution. For instance, the amount of Pt^{4+} used in reactions was 1.2 wt% in TiO_2-based materials and 2.9 wt% for bismuth and copper compounds in Figure 3. It can also be seen that the amount of Pt particles on bismuth and copper particles were much higher than those on TiO_2 particles. This result could correspond well with the added amount of $PtCl_4$ solution in the reaction system. Therefore, it can be inferred that the 1.2 wt% Pt^{4+} loading amount is not enough for TiO_2 nanomaterials. In other words, more Pt nanoparticles would be deposited on the surface of TiO_2 nanomaterials if more $PtCl_4$ solution was added during the reaction.

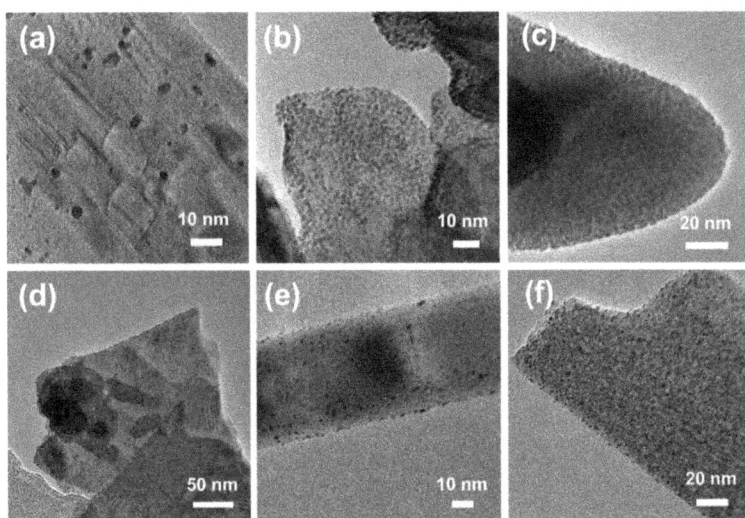

Figure 3. TEM images of Pt-deposited TiO$_2$ MP (**a**), Bi$_2$WO$_6$ (**b**), BiVO$_4$ (**c**), CuS (**d**), TiO$_2$ t1 (**e**), and TiO$_2$ t2 (**f**).

To further verify the existence of the loaded Pt particles, SAED and EDX analyses were applied on the Pt-deposited TiO$_2$ nanofibers (Figure 4). In Figure 4b, a pair of diffraction spots, labelled by a red circle with a d-spacing of 2.26 Å, was observed. This diffraction data was correlated well with (111) crystal plane of Pt, which finally confirmed the presence of Pt in this Pt-deposited TiO$_2$ nanomaterial. Additionally, the scanning TEM (STEM), and the corresponding EDX mapping signal, authenticate uniform distribution of elemental Ti and O throughout the TiO$_2$ nanofibers, as well as random distribution of the elemental Pt on the fiber's surface.

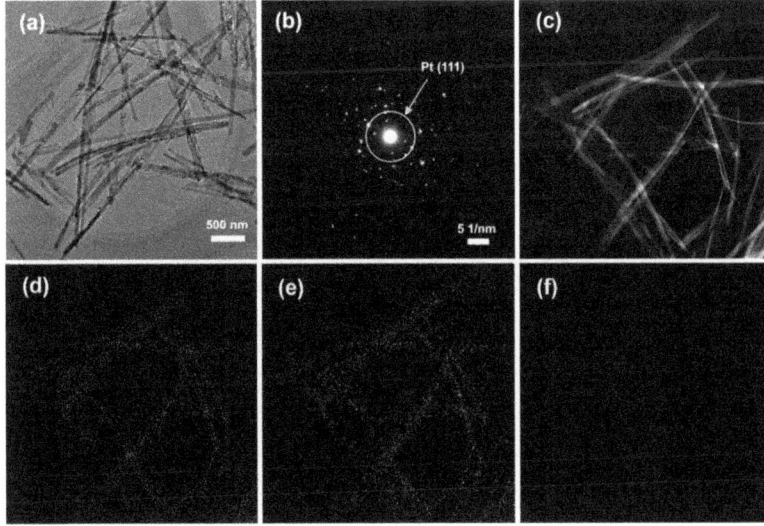

Figure 4. TEM image of Pt deposited TiO$_2$ MP (**a**) and the corresponding SAED patterns (**b**); STEM image of Pt deposited TiO$_2$ MP (**c**), and the corresponding EDX mapping images of elemental Ti (**d**), O (**e**), and Pt (**f**).

Above results revealed the successful deposition of Pt nanoparticles on various photocatalyst nanomaterials, including TiO_2-based, bismuth-based, and copper-based materials; the Pt particles almost occupied the entire surface of core particles. The TEM results also indicated that the Pt particles did not show any selectivity on deposition position; they deposited very randomly. This is considered to be the result of a surface defects-driven electron escape mechanism. In our experiments, only photocatalyst nanoparticles were presented together with the Pt source ($PtCl_4$) in aqueous. Under the illumination, electrons generated and migrated to the particle surface to react with adsorbed Pt^{4+} ions. Pt^{4+} ions should be reduced at the electron escaping sites, resulting in the formation of Pt nanoparticle-deposited photocatalyst nanomaterials. As mentioned before, surface defects played important roles for electron migration and escape in monodispersed semiconductor nanoparticles. In general, defects distributed on the surface randomly. This information explained that Pt nanoparticles randomly deposited on semiconductor nanoparticles, which confirmed that the Pt deposition is a result of surface defects engineering. Meanwhile, with different amounts of $PtCl_4$ used in the reaction, the amount of Pt nanoparticles deposited on nanofibers changed. As shown in Table 2, the Pt^{4+} loading amount increased from 1.2 wt% to 12.0 wt%, and the Pt-deposited amount increased, linearly, from 0.48 wt% to 7.97 wt%. This could also be evidence that the Pt deposition on monodisperse nanoparticles mainly occurred due the surface defects-driven electron escape mechanism.

Table 2. Summary of Pt^{4+} loading amounts and actual Pt-deposited amounts, analyzed by EDX, for TiO_2-based materials.

Basic Materials	Pt^{4+} Loading Amount	Pt Deposited Amount (Analyzed by EDX)
TiO_2 t1	1.2 wt%	0.76 wt%
	2.3 wt%	2.05 wt%
	5.8 wt%	4.07 wt%
	12.0 wt%	7.96 wt%
TiO_2 t2	1.2 wt%	0.31 wt%
	2.3 wt%	3.15 wt%
	5.8 wt%	6.03 wt%
	12.0 wt%	8.70 wt%
TiO_2 MP	1.2 wt%	0.37 wt%
	2.3 wt%	2.30 wt%
	5.8 wt%	4.20 wt%
	12.0 wt%	7.26 wt%

In order to understand the reaction mechanism well, comparison experiments and replenish analyses have been done. Firstly, besides the studies on the precipitate phase, Pt deposited TiO_2 nanomaterials during the solution phase, which was supposed to contain excess $PtCl_4$ and some small photocatalyst nanoparticles, has also been investigated by TEM and EDX analyses. TEM results revealed that there are no free Pt nanoparticles observed, while the EDX mapping signal of Pt has only been detected on residual TiO_2 nanofibers. Secondly, a comparison experiment, based on a TiO_2 MP nanofiber sample, was taken in the dark. EDX data shows 0.98 wt% Pt signal. However, the sample prepared with the same condition as the comparison experiment showed a 7.26 wt% Pt signal after illumination for 30 min. Combined with TEM and SAED results, there are neither small particles vision nor a Pt diffraction signal (a d-spacing of 2.26 Å). Such a small amount of Pt loading is regarded as Pt^{4+} ions adsorbed on particle surface. These comparison experiment results further prove the reaction mechanism: Pt^{4+} ions adsorbed on a particle's surface and reacted with escaped electrons to form Pt nanoparticles, which means Pt nanoparticles were formed and deposited, directly on the particle surface, at electron escaping sites. This mechanism makes it possible to identify electron escaping conditions, directly, by the visual detection of Pt nanoparticles.

Another part of our work was the study on heterostructured TiO_2 MP nanofibers. As verified by XRD analysis, the TiO_2 MP sample contained both brookite and anatase phases. The HRTEM images, as shown in Figure 5, also revealed the co-existence of the two phases' particles in this sample. In the HRTEM picture of the left nanofiber (Figure 5b), two lattice fringes, with d spacing 6.26 Å and 3.69 Å, were correlated well with that of the (001) and (201) planes of the brookite phase (JCPDS No. 46-1238), respectively. For the right counterpart (Figure 5c), lattice fringes, with a d spacing 3.54 Å, could be indexed to the (101) planes of the anatase phase (JCPDS #21-1272). The HRTEM results were in accordance with the XRD analysis.

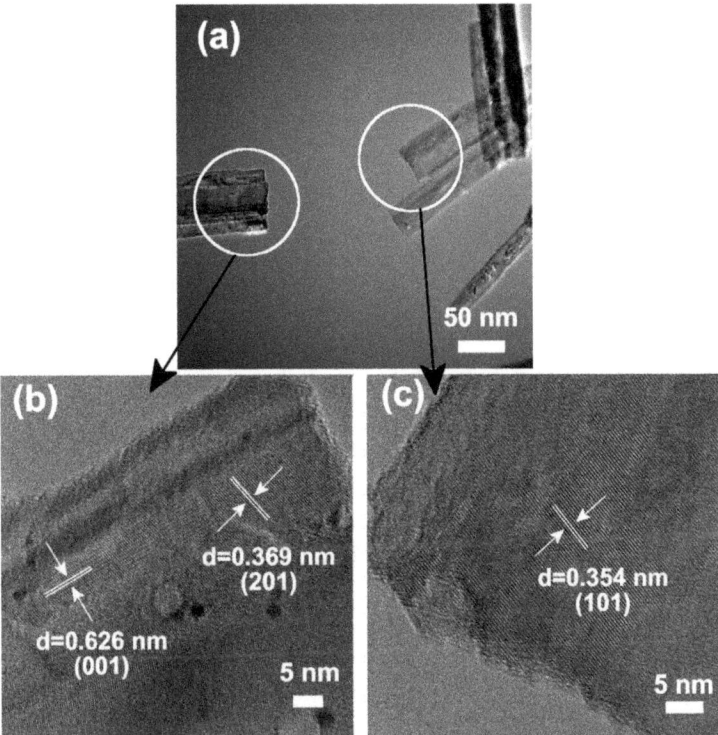

Figure 5. (a) TEM images of Pt-deposited TiO_2 MP sample (with 1.2 wt% Pt^{4+} loading amount). HRTEM images of (b) brookite TiO_2 nanofibers and (c) anatase TiO_2 nanofibers.

Interestingly, the Pt nanoparticles preferred to be reduced on the brookite TiO_2 nanofibers instead of the anatase TiO_2 nanofibers, which could be seen from Figure 5a. As analyzed in the above TEM results (Figure 3e,f), in pure TiO_2 t1 and t2 samples, most of the particles show obvious Pt deposition in both brookite and anatase TiO_2 nanofibers when 1.2 wt% $PtCl_4$ was added. Considering that the Pt nanoparticles were only photodeposited at the electron escape site, the electron escape sites should exist in both brookite and anatase TiO_2 nanofibers. However, when the same amount of $PtCl_4$ was used to react with TiO_2 MP nanoparticles, a different phenomenon was observed. In the mixed phase sample, Pt nanoparticles were deposited on partial nanofibers. By analyzing TEM, it seems that Pt nanoparticles were preferred to deposit on brookite phase nanofibers (Figure 5b) compared to anatase fibers (Figure 5c), when the Pt^{4+} loading amount is 1.2 wt%. This can also be evidenced by the EDX analysis data in Table 2. Compared to the pure t1 and t2 samples, mixed phase TiO_2 nanofibers show almost the same Pt deposition amount. This means the Pt nanoparticles were mainly deposited on one kind of TiO_2 nanomaterial instead of not being deposited. Furthermore, this phenomenon became more obvious when the Pt^{4+} loading amount decreased to 0.70 wt% and even to 0.23 wt%. By studying a relatively low Pt^{4+} loading amount sample, we were able to distinguish these particles more evidently, with or without loading Pt particles. It confirmed that nanofibers with the brookite phase showed higher reaction activities for the photocatalytic metal deposition compared to anatase phase particles. This can be attributed to the difference of work function. Referring to Vera and co-workers' results [37], work function of brookite TiO_2 is a little bit smaller than in the anatase phase. Although work function value is variable for different testing methods, in different testing environments, or even based on different facets, work function of the brookite phase is always smaller than anatase under the same conditions. EDX analysis results also present the same trend, in which pure TiO_2 t2 samples have a higher Pt deposited amount.

Except for the two kinds of pure phase particles, a two phase co-existence structure was fabricated side by side on one single particle. In Figure 6, a heterostructure TiO_2 nanofiber, with brookite and anatase phases, was observed. It is obvious that a distinct interface between the two phases was observed, in one particle, by the bright-field TEM image. Figure 6b showed the high resolution image of the upper side, in which the lattice fringes, with a d spacing 2.31 Å, was identified. This area was correlated with (112) the crystal plane of the anatase TiO_2 phase. The bottom part, with d spacing 5.88 Å lattice fringes, has been decided as the brookite phase. TEM results showed that the two phases co-existed, and a special heterostructure containing the brookite and anatase phases was constructed in one particle.

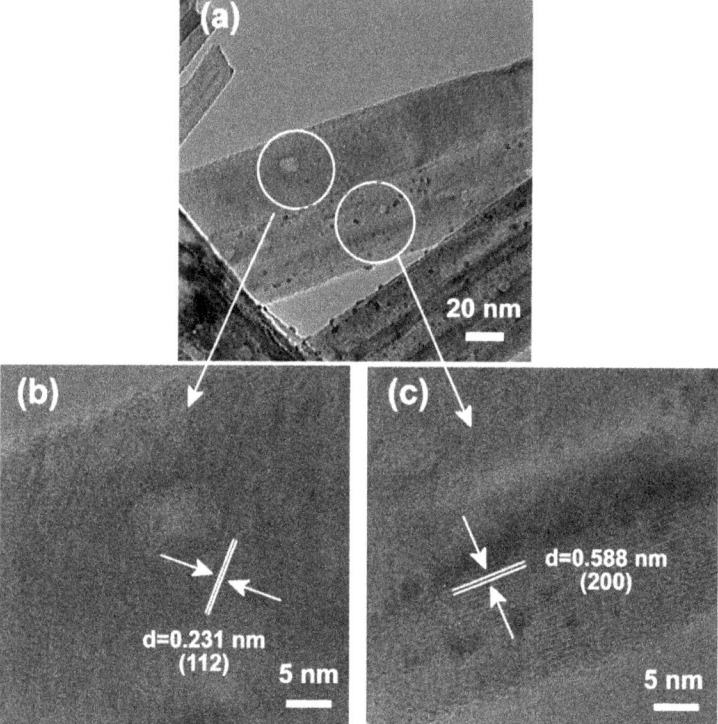

Figure 6. TEM images of a nanoparticle with a heterostructure (**a**) in the Pt-deposited TiO_2 MP sample (with 1.2 wt% Pt^{4+} loading amount). HRTEM images of (**b**) upper and (**c**) lower areas, as indicated in (**a**).

Meanwhile, TEM results also revealed that almost all the Pt nanoparticles are loaded on the bottom part, which is the brookite phase. This phenomenon declared that the electrons mainly escaped from the surface of the brookite TiO_2. As discussed previously, there are two effects that could be used to explain this phenomenon. First, some heterostructure materials have an electrical junction that could promote the separation of the charge carrier and drive the electrons across the junction. In this mixed phase sample, the electrons mainly escaped from the brookite phase. Thus, it is inferred that the brookite phase was performed as an electron attractor component. The existence of heterojunction efficiently causes the migration of electrons across the junction interface, and then, they come out from the surface of the counterpart with the electron attractor component. It resulted in the higher content of escaped electrons from brookite TiO_2 particles compared with the pure brookite TiO_2 sample. Therefore, the Pt nanoparticles, attached on the brookite TiO_2 particles in the heterojunction sample, were more than that in the pure brookite TiO_2 sample. Second, the different work functions of these two phases would be other driving forces for the different Pt deposition densities.

Figure 7 shows the TEM images of Pt-deposited TiO_2 MP nanofibers with a Pt^{4+} loading amount from 2.3 wt% to 12.0 wt%. From Figure 7a,b, a clear viewer of the different Pt deposited densities, on two sides of the heterostructures, was observed even if the Pt^{4+} loading amount increased to 5.8 wt%. The work functions, different between brookite phase (around 3.72 eV) and anatase phase (around 3.84 eV) [37], may not be big enough to result in such different Pt-deposited densities, especial at higher Pt^{4+} loading amounts. Thus, we believe that both the heterojunction driven force and work function difference contribute to the Pt deposited density difference between brookite and anatase phases in heterostructured nanoparticles. In Figure 7c, Pt nanoparticles abundantly deposited on both sides, which could be attributed to the saturation of active sites on the brookite component surface.

Figure 7. TEM images of Pt-deposited TiO_2 MP nanofibers with Pt^{4+} loading amount 2.3 wt% (**a**), 5.8 wt% (**b**), and 12.0 wt% (**c**), respectively.

From another point of view, the Pt deposit sites directly point out the electron escape site. In monodisperse nanomaterials, Pt nanoparticles are loaded randomly. This attributes to the abundant surface defects of photocatalyst materials. Surface defects are considered as the main electron escape driving force. Different from this, in some heterostructure materials, electrons are driven to migrate to the electron attractor component, due to the existence of heterojunction structure and the difference of work functions. Thus, although Pt^{4+} ions are already adsorbed on the entire surface, only those ions located at the electron escape side are successfully reduced. Our results correspond well with those.

4. Conclusions

In summary, electron escaping conditions of various kinds of photocatalyst materials have been studied through photodeposition reactions. For photocatalyst nanomaterials, which show monodisperse crystal structure, Pt nanoparticles deposited randomly on the surface. This is attributed to the rich surface defects. In the heterostructured photocatalyst nanomaterials, Pt nanoparticles are preferred to load at one side. This is due to the electron migration driving force provided by heterojunction structure and work function difference. This is considered as a higher force, to promote the separation and migration of electrons, than the one caused by surface defects. Our results correspond well with known conclusions published previously. This is enough to demonstrate that photodeposition is usable for the study of electron escaping conditions. Further, it can be used as a "tool" to identify the catalytic ability of photocatalyst materials—especially the activity of their single particle.

Author Contributions: Conceptualization, C.Y. and Y.H.; methodology, C.Y.; formal analysis, C.Y. and Y.H.; writing—original draft preparation, C.Y.; writing—review and editing, Y.H. All authors have read and agreed to the published version of the manuscript.

Funding: This work was supported by the Project of Shandong Province Higher Educational Science and Technology Program (No. J18KA073), the Doctoral Foundation of University of Jinan (No. XBS1643), the National Natural Science Foundation of China (grants nos. 52072150 and 51972146), and the China Association for Science and Technology (Young Elite Scientists Sponsorship Program).

Institutional Review Board Statement: Not applicable.

Informed Consent Statement: Not applicable.

Data Availability Statement: The data presented in this study are available on request from the corresponding author.

Conflicts of Interest: There are no conflict to declare.

References

1. Fujishima, A.; Honda, K. Electrochemical Photolysis of Water at a Semiconductor Electrode. *Nature* **1972**, *238*, 37–38. [CrossRef] [PubMed]
2. Fick, J. Chapter 8—Crystalline Nanoparticles in Glasses for Optical Applications. In *Handbook of Surfaces and Interfaces of Materials*; Nalwa, H.S., Ed.; Academic Press: Burlington, VT, USA, 2001; pp. 311–350.
3. Clarizia, L.; Russo, D.; Di Somma, I.; Andreozzi, R.; Marotta, R. 9—Metal-based semiconductor nanomaterials for photocatalysis. In *Multifunctional Photocatalytic Materials for Energy*; Lin, Z., Ye, M., Wang, M., Eds.; Elsevier Ltd.: Amsterdam, The Netherlands, 2018; pp. 187–213.
4. Ye, C.; Li, Z.; Ye, E. Chapter 4 Metal–Oxide Semiconductor Nanomaterials for Photothermal Catalysis. In *Photothermal Nanomaterials*; The Royal Society of Chemistry: Cambridge, UK, 2022; pp. 135–157.
5. Regulacio, M.D. Chapter 5 Copper Sulfide-based Nanomaterials for Photothermal Applications. In *Photothermal Nanomaterials*; The Royal Society of Chemistry: Cambridge, UK, 2022; pp. 158–185.
6. Serpone, N.; Emeline, A.V. Semiconductor Photocatalysis—Past, Present, and Future Outlook. *J. Phys. Chem. Lett.* **2012**, *3*, 673–677. [CrossRef] [PubMed]
7. Mohd Kaus, N.H.; Rithwan, A.F.; Adnan, R.; Ibrahim, M.L.; Thongmee, S.; Mohd Yusoff, S.F. Effective Strategies, Mechanisms, and Photocatalytic Efficiency of Semiconductor Nanomaterials Incorporating rGO for Environmental Contaminant Degradation. *Catalysts* **2021**, *11*, 302. [CrossRef]
8. Bai, S.; Jiang, W.; Li, Z.; Xiong, Y. Surface and Interface Engineering in Photocatalysis. *ChemNanoMat* **2015**, *1*, 223–239. [CrossRef]

9. Li, J.; Lou, Z.; Li, B. Engineering plasmonic semiconductors for enhanced photocatalysis. *J. Mater. Chem. A* **2021**, *9*, 18818–18835. [CrossRef]
10. Ren, H.; Yang, J.-L.; Yang, W.-M.; Zhong, H.-L.; Lin, J.-S.; Radjenovic, P.M.; Sun, L.; Zhang, H.; Xu, J.; Tian, Z.-Q.; et al. Core–Shell–Satellite Plasmonic Photocatalyst for Broad-Spectrum Photocatalytic Water Splitting. *ACS Mater. Lett.* **2021**, *3*, 69–76. [CrossRef]
11. Su, H.; Wang, W. Dynamically Monitoring the Photodeposition of Single Cocatalyst Nanoparticles on Semiconductors via Fluorescence Imaging. *Anal. Chem.* **2021**, *93*, 11915–11919. [CrossRef]
12. Fu, Y.-S.; Li, J.; Li, J. Metal/Semiconductor Nanocomposites for Photocatalysis: Fundamentals, Structures, Applications and Properties. *Nanomaterials* **2019**, *9*, 359. [CrossRef]
13. Vinasco, J.A.; Radu, A.; Kasapoglu, E.; Restrepo, R.L.; Morales, A.L.; Feddi, E.; Mora-Ramos, M.E.; Duque, C.A. Effects of Geometry on the Electronic Properties of Semiconductor Elliptical Quantum Rings. *Sci. Rep.* **2018**, *8*, 13299. [CrossRef]
14. Tavkhelidze, A.; Bibilashvili, A.; Jangidze, L.; Gorji, N.E. Fermi-Level Tuning of G-Doped Layers. *Nanomaterials* **2021**, *11*, 505. [CrossRef]
15. Koochi, H.; Ebrahimi, F. Geometrical effects on the electron residence time in semiconductor nano-particles. *J. Chem. Phys.* **2014**, *141*, 094702. [CrossRef]
16. Hemmerling, J.R.; Mathur, A.; Linic, S. Characterizing the Geometry and Quantifying the Impact of Nanoscopic Electrocatalyst/Semiconductor Interfaces under Solar Water Splitting Conditions. *Adv. Energy Mater.* **2022**, 2103798. [CrossRef]
17. Raizada, P.; Soni, V.; Kumar, A.; Singh, P.; Parwaz Khan, A.A.; Asiri, A.M.; Thakur, V.K.; Nguyen, V.-H. Surface defect engineering of metal oxides photocatalyst for energy application and water treatment. *J. Mater.* **2021**, *7*, 388–418. [CrossRef]
18. Yang, S.; Zhu, X.; Peng, S. Prospect Prediction of Terminal Clean Power Consumption in China via LSSVM Algorithm Based on Improved Evolutionary Game Theory. *Energies* **2020**, *13*, 2065. [CrossRef]
19. Lu, Q.; Yu, Y.; Ma, Q.; Chen, B.; Zhang, H. 2D Transition-Metal-Dichalcogenide-Nanosheet-Based Composites for Photocatalytic and Electrocatalytic Hydrogen Evolution Reactions. *Adv. Mater.* **2016**, *28*, 1917–1933. [CrossRef]
20. Ma, F.; Wu, Y.; Shao, Y.; Zhong, Y.; Lv, J.; Hao, X. 0D/2D nanocomposite visible light photocatalyst for highly stable and efficient hydrogen generation via recrystallization of CdS on MoS2 nanosheets. *Nano Energy* **2016**, *27*, 466–474. [CrossRef]
21. Li, C.; Yu, S.; Dong, H.; Liu, C.; Wu, H.; Che, H.; Chen, G. Z-scheme mesoporous photocatalyst constructed by modification of Sn_3O_4 nanoclusters on $g-C_3N_4$ nanosheets with improved photocatalytic performance and mechanism insight. *Appl. Catal. B* **2018**, *238*, 284–293. [CrossRef]
22. Zafar, Z.; Yi, S.; Li, J.; Li, C.; Zhu, Y.; Zada, A.; Yao, W.; Liu, Z.; Yue, X. Recent Development in Defects Engineered Photocatalysts: An Overview of the Experimental and Theoretical Strategies. *Energy Environ. Mater.* **2022**, *5*, 68–114. [CrossRef]
23. Li, H.; Li, J.; Ai, Z.; Jia, F.; Zhang, L. Oxygen Vacancy-Mediated Photocatalysis of BiOCl: Reactivity, Selectivity, and Perspectives. *Angew. Chem. Int. Ed.* **2018**, *57*, 122–138. [CrossRef]
24. Bai, S.; Zhang, N.; Gao, C.; Xiong, Y. Defect engineering in photocatalytic materials. *Nano Energy* **2018**, *53*, 296–336. [CrossRef]
25. Hu, J.; Zhao, X.; Chen, W.; Chen, Z. Enhanced Charge Transport and Increased Active Sites on α-Fe_2O_3 (110) Nanorod Surface Containing Oxygen Vacancies for Improved Solar Water Oxidation Performance. *ACS Omega* **2018**, *3*, 14973–14980. [CrossRef] [PubMed]
26. Zhang, W.; Song, L.; Cen, J.; Liu, M. Mechanistic Insights into Defect-Assisted Carrier Transport in Bismuth Vanadate Photoanodes. *J. Phys. Chem. C* **2019**, *123*, 20730–20736. [CrossRef]
27. Zhou, M.; Lou, X.W.; Xie, Y. Two-dimensional nanosheets for photoelectrochemical water splitting: Possibilities and opportunities. *Nano Today* **2013**, *8*, 598–618. [CrossRef]
28. Tan, H.; Zhao, Z.; Zhu, W.-b.; Coker, E.N.; Li, B.; Zheng, M.; Yu, W.; Fan, H.; Sun, Z. Oxygen Vacancy Enhanced Photocatalytic Activity of Pervoskite $SrTiO_3$. *ACS Appl. Mater. Interfaces* **2014**, *6*, 19184–19190. [CrossRef]
29. Wang, G.; Huang, B.; Li, Z.; Lou, Z.; Wang, Z.; Dai, Y.; Whangbo, M.-H. Synthesis and characterization of ZnS with controlled amount of S vacancies for photocatalytic H_2 production under visible light. *Sci. Rep.* **2015**, *5*, 8544. [CrossRef]
30. Cui, H.; Liu, H.; Shi, J.; Wang, C. Function of TiO_2 Lattice Defects toward Photocatalytic Processes: View of Electronic Driven Force. *Int. J. Photoenergy* **2013**, *2013*, 364802. [CrossRef]
31. Pan, L.; Wang, S.; Xie, J.; Wang, L.; Zhang, X.; Zou, J.-J. Constructing TiO_2 p-n homojunction for photoelectrochemical and photocatalytic hydrogen generation. *Nano Energy* **2016**, *28*, 296–303. [CrossRef]
32. Chen, Y.; Crittenden, J.C.; Hackney, S.; Sutter, L.; Hand, D.W. Preparation of a Novel TiO_2-Based p−n Junction Nanotube Photocatalyst. *Environ. Sci. Technol.* **2005**, *39*, 1201–1208. [CrossRef]
33. Wang, M.; Hu, Y.; Han, J.; Guo, R.; Xiong, H.; Yin, Y. TiO_2/NiO hybrid shells: P–n junction photocatalysts with enhanced activity under visible light. *J. Mater. Chem. A* **2015**, *3*, 20727–20735. [CrossRef]
34. Lee, Y.; Kim, E.; Park, Y.; Kim, J.; Ryu, W.; Rho, J.; Kim, K. Photodeposited metal-semiconductor nanocomposites and their applications. *J. Mater.* **2018**, *4*, 83–94. [CrossRef]
35. Yu, J.; Qi, L.; Jaroniec, M. Hydrogen Production by Photocatalytic Water Splitting over Pt/TiO_2 Nanosheets with Exposed (001) Facets. *J. Phys. Chem. C* **2010**, *114*, 13118–13125. [CrossRef]

36. Li, Q. Collagen Deposition and Fibrosis in the Lymphatic Tissues of HIV-1 Infected Individuals. In *Encyclopedia of AIDS*; Hope, T.J., Stevenson, M., Richman, D., Eds.; Springer: New York, NY, USA, 2021; pp. 1–6.
37. Mansfeldova, V.; Zlamalova, M.; Tarabkova, H.; Janda, P.; Vorokhta, M.; Piliai, L.; Kavan, L. Work Function of TiO_2 (Anatase, Rutile, and Brookite) Single Crystals: Effects of the Environment. *J. Phys. Chem. C* **2021**, *125*, 1902–1912. [CrossRef]

Article

Characterization and Microstructural Evolution of Continuous BN Ceramic Fibers Containing Amorphous Silicon Nitride

Yang Li [1,2], Min Ge [1,2], Shouquan Yu [1,3], Huifeng Zhang [1,3], Chuanbing Huang [1,3], Weijia Kong [1], Zhiguang Wang [1,2] and Weigang Zhang [1,2,3,*]

[1] Key Laboratory of Science and Technology on Particle Materials, Key Laboratory of Multiphase Complex Systems, Institute of Process Engineering, Chinese Academy of Sciences, Beijing 100190, China; liyang19@ipe.ac.cn (Y.L.); gemin@ipe.ac.cn (M.G.); sqyu@ipe.ac.cn (S.Y.); hfzhang@ipe.ac.cn (H.Z.); cbhuang@ipe.ac.cn (C.H.); wjkong19@ipe.ac.cn (W.K.); wangzhiguang18@mails.ucas.ac.cn (Z.W.)
[2] School of Chemical Engineering, University of Chinese Academy of Sciences, Beijing 100049, China
[3] School of Rare Earth, University of Science and Technology of China, Ganzhou 341000, China
* Correspondence: wgzhang@ipe.ac.cn; Tel./Fax: +86-10-8254-4908

Abstract: Boron nitride (BN) ceramic fibers containing amounts of silicon nitride (Si_3N_4) were prepared using hybrid precursors of poly(tri(methylamino)borazine) (PBN) and polycarbosilane (PCS) via melt-spinning, curing, decarburization under NH_3 to 1000 °C and pyrolysis up to 1600 °C under N_2. The effect of Si_3N_4 contents on the microstructure of the BN/Si_3N_4 composite ceramics was investigated. Series of the BN/Si_3N_4 composite fibers containing various amounts of Si_3N_4 from 5 wt% to 25 wt% were fabricated. It was found that the crystallization of Si_3N_4 could be totally restrained when its content was below 25 wt% in the BN/Si_3N_4 composite ceramics at 1600 °C, and the amorphous BN/Si_3N_4 composite ceramic could be obtained with a certain ratio. The mean tensile strength and Young's modulus of the composite fibers correlated positively with the Si_3N_4 mass content, while an obvious BN (shell)/Si_3N_4 (core) was formed only when the Si_3N_4 content reached 25 wt%.

Keywords: composite ceramic fibers; boron nitride; silicon nitride

1. Introduction

Advanced antenna radomes or windows are widely used in aerospace radar devices, and wave-transparent materials with high performance are urgently required [1–3]. Continuous ceramic fiber reinforced ceramic matrix composites are the most promising materials, and several inorganic continuous fibers, such as quartz, Si_3N_4 and BN fibers, are suitable candidates as reinforcements [4,5]. Because of the occurrence of re-crystallization and particle coarsening, quartz fibers undergo serious degradation of mechanical strength at temperatures above 900 °C [6]. Although BN fibers show excellent high-temperature stability, with a melting point above 2900 °C [7–9], their low mechanical strength impedes any further development [10,11]. Generally, Si_3N_4 fibers possess higher mechanical strength than BN fibers, while the thermal resistance and the dielectric properties of Si_3N_4 fibers are inferior to BN fibers [12]. Therefore, it is urgent to develop new types of ceramic fibers containing B, N, and Si to combine the merits of both BN and Si_3N_4 fibers in order to ensure excellent dielectric properties, high mechanical properties and good thermal resistance [13–16].

The polymer-derived ceramics (PDCs) method is the only feasible approach to prepare SiBN fibers; it can be used to design the atomic composition and the microstructure of fibers with low impurity [17–20]. Numerous researchers have synthesized different types of precursors for SiBN ceramic fibers. Tang et al. [21] obtained SiBN fibers using a novel precursor fabricated by the reaction of boron trichloride, dichloromethylsilane and hexamethyldisilazane and the obtained SiBN fibers had excellent mechanical strength up to 1.83 GPa as well as thermally stable dielectric properties. Liu et al. [22] prepared SiBN ternary ceramic fibers with a tensile strength of 0.87 GPa from a polymer precursor made from the reaction of hexamethyldisilazane, trichlorosilane, boron trichloride and methylamine. Peng et al. [23] prepared a precursor for SiBN ceramic fibers from the reaction of hexamethyldisilazane, silicon tetrachloride, boron trichloride and methylamine. In summary, a substantial amount of research has been focused on the fabrication of the spinnable polymer precursor containing the Si-N-B bridge bond, since the atomic ratio of Si:B:N of the preceramic polymer is difficult to adjust owing to its high activity [22]. In addition, the synthesis of single-source precursors usually needs multiple-stage processes, which results in significant yield loss and the unavoidable removal of by-products. Additionally, during the multi-step process, the polymeric intermediates are extremely sensitive to moisture, and the requirement for the large inert environment hampers its industrial applications [23].

In a previous study, Tan et al. [24] successfully synthesized BN/Si_3N_4 composite fibers using a mixture of polymers of poly(tri(methylamino)borizane) (PBN) and polysilazane (PSZ), and the fibers showed a mean tensile strength of 1040 MPa. Difficulties in the industrialization of this process exist mainly due to the fact that the PSZ is very sensitive to moisture. Polycarbosilane (PCS) is an organosilicon polymer with a normal structure—$(CH_3HSiCH_2)_x$—that contains a –Si–C– backbone [25] and has been widely used as the precursor for preparation of silicon carbide and silicon nitride ceramic fibers owing to its good solubility and fusibility [26]. Therefore, we adopted a new strategy to prepare BN/Si_3N_4 ceramic fiber by using hybrid polymers of PBN and PCS. The proportions of BN and Si_3N_4 in the final ceramics could be easily adjusted through changing the ratios of PBN/PCS in hybrid composite polymers. The microstructural evolution of the obtained BN/Si_3N_4 composite ceramics and the properties of the composite fibers were investigated.

2. Experiments

2.1. Materials

PBN was synthesized by the reaction of BCl_3 and NH_2CH_3, with a softening point of 80 ± 2 °C [27]. PCS (main units: $(HSi(CH_3)CH_2)$, $(CH_2Si(CH_3)_2CH_2)$, $(Si(CH_3)_2Si(CH_3)_2)$) with a number-average molecular weight of about 1150 and a softening point of 210 ± 2 °C was purchased from Zhongxing New Material Technology Co. Ltd., Ningbo, China. High purity N_2 (>99.99%, Huanyujinghui Co. Ltd., Beijing, China) and NH_3 (>99.99%, Nanfei Co. Ltd., Beijing, China) gases were used. All reactions were carried out under a dry nitrogen atmosphere.

2.2. Preparation of Composite Polymer

PBN and PCS were dissolved in toluene separately, with appropriate ratios, and mixed at 60 °C in a rotary evaporator (Shanghaiyukangkejiaoyiqishebei Co. Ltd., Shanghai, China) for 4 h to form homogeneous hybrid precursor solutions. Then, the solutions were dried at 80 °C for 2 h to remove the toluene. After cooling to room temperature, yellow transparent bulk solids were obtained. Six composite polymers (P1, P2, P3, P4, P5, P6) with different PBN/PCS ratios were prepared. After heating up to 1000 °C in ammonia at a heating rate of 1 K/min and 1600 °C in N_2 at 2 K/min, the BN/Si_3N_4 composite ceramics with different Si_3N_4 mass contents were produced. The final BN and Si_3N_4 mass contents were calculated according to the ceramic yield of each polymer, which was 33 wt% and 58 wt% for PBN and PCS, respectively. The mass contents of PBN and PCS of each hybrid precursor (named P0–P7) as well as the BN and Si_3N_4 mass contents of the corresponding composite ceramics obtained at 1600 °C are listed in Table 1.

Table 1. Composition of hybrid PBN/PCS precursors and the BN/Si$_3$N$_4$ ceramics derivatives (1600 °C).

Materials	Precursor Composition (wt%)		Ceramics Composition (wt%)	
	PBN	PCS	BN	Si$_3$N$_4$
P0	100	0	100	0
P1	97.2	2.8	95	5
P2	91.1	8.9	85	15
P3	84.4	15.6	75	25
P4	77.0	23.0	65	35
P5	64.3	35.7	50	50
P6	37.5	62.5	25	75
P7	0	100	0	100

2.3. Preparation of Composite Fibers

Polymer green fibers were prepared using a lab-scale melt-spinning apparatus (Paigujingmijixie Co. Ltd., Beijing, China). The composite polymer was first fed into the spinning tube and heated to the spinning temperature (about 130 °C) for 3 h. After the removal of the bubbles and an appropriate viscosity being obtained, the molten hybrid polymer was extruded through a single-capillary spinneret 0.20 mm in diameter by controlling the spinning pressure of N$_2$ (roughly 0.5 MPa), and the unmelts were eliminated by a filter. Then, the extrudate flow was drawn into the filament uniaxially and collected on a spool with an appropriate rotating speed of 10 m/s.

Afterward, the green fibers were cured in ammonia at a heating rate of 0.1 K/min to 300 °C and pyrolyzed up to 1000 °C in flowing ammonia at a heating rate of 1 K/min. Then, the fibers were heated to 1600 °C under flowing N$_2$ at 2 K/min. Finally, white composite ceramic fibers were obtained after cooling to ambient temperature.

2.4. Characterization

X-ray diffraction (XRD) studies were carried out with a PANalytical X'Pert-PRO diffractometer (Eindhoven, The Netherlands) at 2θ = 10–90° with Cu Kα radiation (λ = 0.15406 nm at 40 kV and 40 mA). The chemical bonding states were obtained by X-ray Photoelectron Spectroscopy (XPS, ESCALAB 250Xi) (Thermo Fisher Scientific, Waltham, Massachusetts, USA). The element analysis of silicon and boron was conducted using ICP-OES in a ThermoFisher iCAP6300 spectrometer (Waltham, MA, USA), and the nitrogen, carbon and oxygen contents were measured by a vario EL cube analyzer (elementar, Germany). Fiber morphologies were revealed by scanning electron microscopy (SEM) using a JSM-7001F system (JEOL, Tokyo, Japan). Element distribution along the fiber diameter was measured by an EPMA-1720 (Shimadzu, Japan). Single filament tensile properties were determined using an Instron5944 tensile tester (Norwood, MA, USA) with a gauge length of 25 mm, a load cell of 10 N, and a crosshead speed of 5 mm/min. The mean tensile strength of fibers was calculated based on 25 tested fibers using the Weibull statistic, and the Young's modulus of the fibers was evaluated from the strain-stress curves. The dielectric properties of ceramic fibers determined at 10 GHz were measured by the short-circuited wave guide method using an Agilent HP8722ES vector network analyzer (Santa Clara, CA, USA) at ambient temperature based on the Chinese National Standard GB/T 5597-1999.

3. Results and Discussion

It is vital to study the high-temperature stability of polymer-derived ceramics. First, the microstructural developments of pure BN and Si$_3$N$_4$ pyrolyzed at different temperatures from PBN and PCS were investigated by XRD (Figure 1). For BN (Figure 1a), after being pyrolyzed at 1000 °C, two broad diffraction peaks at 2θ = 26.7° and 41.6° appeared. With the pyrolysis temperature rising from 1000 °C to 1600 °C, these diffraction peaks sharpened a little because of ongoing BN crystallization. At 1600 °C, the diffraction peaks were still broad, and no resolutions of the (100) or (101) doublet were displayed, which indicated the formation of BN nanocrystallines. For Si$_3$N$_4$ (Figure 1b), it can be seen that the as-pyrolyzed ceramics were amorphous below 1400 °C, and the crystallization process started at 1500 °C, which would affect the high temperature stability of ceramics or ceramic fibers.

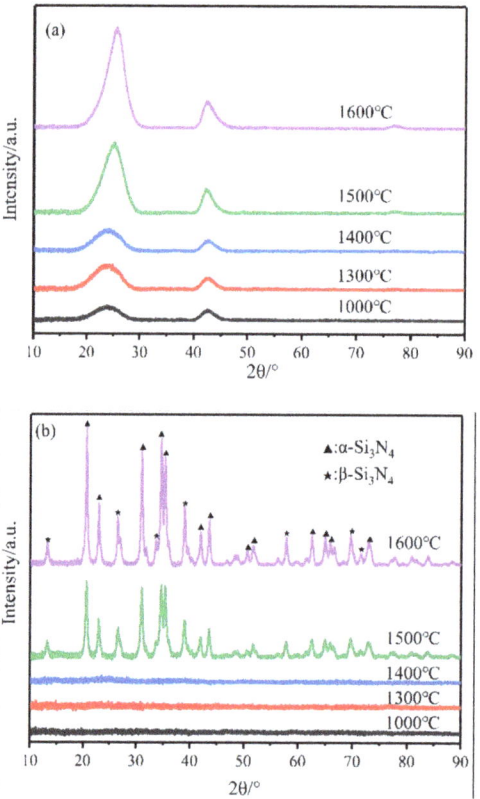

Figure 1. XRD patterns of (**a**) BN and (**b**) Si_3N_4 pyrolyzed at different temperatures from PBN and PCS.

To elucidate microstructural evolution of these composite polymer-derived ceramics, six hybrid precursors with different PBN/PCS ratios were fabricated and pyrolyzed at 1600 °C. The calculative mass contents of BN and Si_3N_4 in the composite ceramics (P0-7-C) are listed in Table 1. The chemical environment of B, N and Si atoms in the composite ceramic (P3-C) was studied by XPS (Figure 2). The B_{1s} peak at 190.8 eV and the N_{1s} at 398.1 eV confirmed the presence of BN. Moreover, the binding energy centered at 102.5 eV for Si_{2p} and 399.7 eV for N_{1s} demonstrated the existence of Si–N bonds, indicating that the composite ceramic was composed of a mixture of BN and Si_3N_4.

The elemental compositions of these polymer-derived composite ceramics obtained at 1600 °C (P0–7-C) are listed in Table 2. Additionally, XRD patterns of these composite ceramics are shown in Figure 3. Obviously, when the content of the BN was over 75 wt% in the composite ceramics, no silicon nitride peaks were detected, indicating that the BN phase restrained the decomposition of Si_3N_4 and limited the grain size of Si_3N_4 crystals. Likewise, with the increase of Si_3N_4 mass content in these composite ceramics, the crystallization process of BN was also hindered, and there existed no BN peaks when the mass content of Si_3N_4 exceeded 15 wt%. Noticeably, the P3-C containing 75 wt% BN and 25 wt% Si_3N_4 was totally amorphous, which indicated that the clusters of BN and Si_3N_4 totally hindered the crystallization of each other. Its amorphous state at 1600 °C would guarantee reliable performance for the composite ceramics or ceramic fibers in high-temperature environments. However, unlike the results of Tan et al. [14], who used PSZ as the raw materials rather than PCS, no h-BN crystals were found in this system. After analyzing the oxygen contents of PSZ and PCS, which were 3% and 0.6%, respectively, it was concluded that the introduction of oxygen could promote the crystallization process of h-BN by formulating the oxides with low melting points.

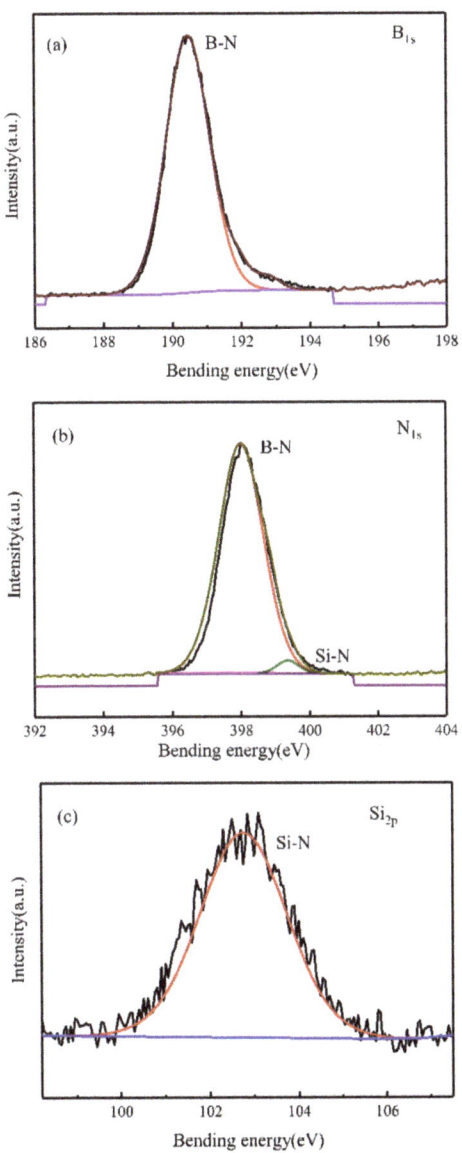

Figure 2. XPS spectra of the composite ceramic (P3-C): (**a**) B_{1s}; (**b**) N_{1s}; (**c**) Si_{2p}.

Table 2. Elemental content of different BN/Si3N4 composite ceramics.

Elemental Content (wt%)	Si	B	N	C
P0-C	0	43.7	56.1	0.2
P1-C	3.4	41.3	55.2	0.1
P2-C	8.7	36.9	54.2	0.2
P3-C	15.6	32.6	51.7	0.1
P4-C	19.7	29.3	50.8	0.2
P5-C	29.5	21.7	48.7	0.1
P6-C	43.2	10.4	44.3	0.3
P7-C	60.1	0	39.8	0.1

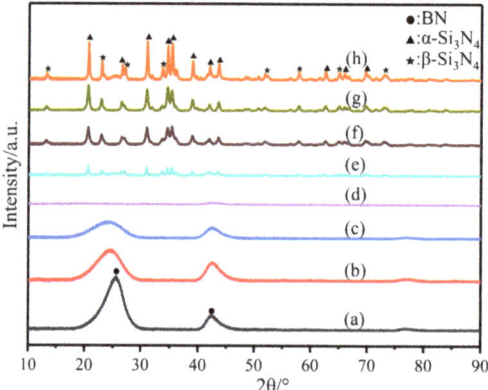

Figure 3. XRD patterns of the composite ceramics pyrolyzed at 1600 °C: (**a**) P0-C; (**b**) P1-C; (**c**) P2-C; (**d**) P3-C; (**e**) P4-C; (**f**) P5-C; (**g**) P6-C; (**h**) P7-C.

The composite ceramics pyrolyzed at 1600 °C were further annealed at 1700 °C under N_2 for 2 h, and the corresponding XRD patterns are shown in Figure 4. Except for P1-C, the composite ceramics all exhibited Si_3N_4 crystals, which was not beneficial for the high-temperature stability of the composite ceramics. In summary, the BN/Si_3N_4 composite ceramics could only keep stability below 1700 °C with appropriate ratios.

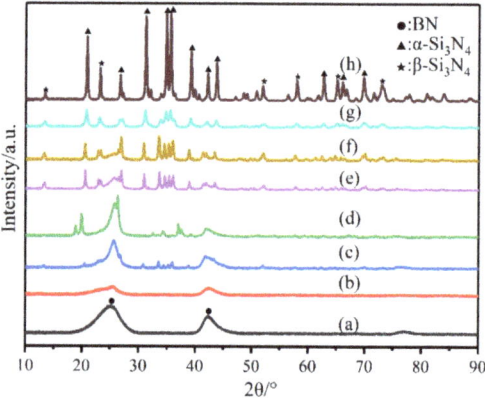

Figure 4. XRD patterns of the composite ceramics pyrolyzed at 1700 °C: (**a**) P0-C; (**b**) P1-C; (**c**) P2-C; (**d**) P3-C; (**e**) P4-C; (**f**) P5-C; (**g**) P6-C; (**h**) P7-C.

Based on the investigation of the microstructural evolution of BN/Si$_3$N$_4$ composite ceramics, the crystallization of Si$_3$N$_4$ could be totally restrained when its content in the composite ceramics was below 25% at 1600 °C. Then, the composite ceramic fibers were fabricated from hybrid precursor P1, P2 and P3 through melt-spinning, curing and decarburization in NH$_3$ under 1000 °C and pyrolysis at 1600 °C in N$_2$. The elemental compositions of these obtained fibers (P1-F, P2-F and P3-F) are listed in Table 3. Additionally, the XRD spectra of these fibers are shown in Figure 5. All these fibers only showed two broad diffuse peaks, revealing the low crystallinity of BN [28,29]. Noticeably, no diffraction peaks of Si$_3$N$_4$ in the composite fibers were detected, indicating that Si$_3$N$_4$ existed in an amorphous state, which was beneficial to the high temperature stability of the composite fibers.

Table 3. Elemental content of BN/Si$_3$N$_4$ composite fibers.

Elemental Content (wt%)	Si	B	N	C
P1-F	3.2	41.3	55.3	0.2
P2-F	8.2	37.2	54.5	0.1
P3-F	14.6	33.1	52.3	0.1

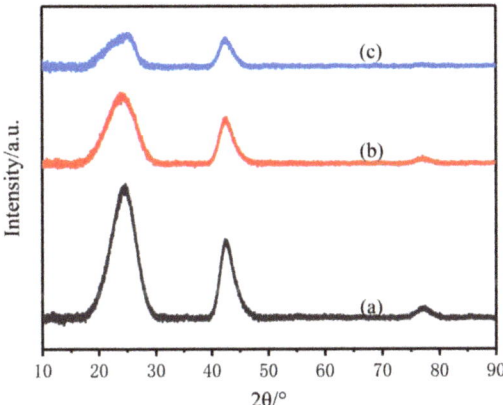

Figure 5. XRD patterns of the composite ceramic fibers: (**a**) P1-F; (**b**) P2-F; (**c**) P3-F.

Figure 6 shows the morphologies of the obtained BN/Si$_3$N$_4$ composite fibers pyrolyzed at 1600 °C. The diameter of the fibers was roughly 12 μm, and the surface was smooth and compact, without any apparent voids. The cross sections were nearly circular without inter-fusion, exhibiting a glass-like fracture feature, which demonstrated that the curing and pyrolysis process could meet the preparation requirements.

In order to clarify the distributions of these elements (Si, B, N), the fibers were embedded in epoxy resin, with further polishing and spraying carbon, and then characterized by EPMA (Figure 7). Obviously, the distributions of each atom for P1-F and P2-F were nearly homogeneous. For P3-F, the atomic Si aggregated in the core, while the concentration of atomic B was higher in the outside shell, revealing the phase separation of Si$_3$N$_4$ and BN. It was concluded that during the spinning process, the PCS tended to aggregate in the core under the shearing pressure owing to the huge differences of the viscosity and softening point of PBN (80 °C) and PCS (210 °C), which led to the unique structure of the final composite fiber; this structure could only be obtained when the Si$_3$N$_4$ content of the fibers reached 25 wt%.

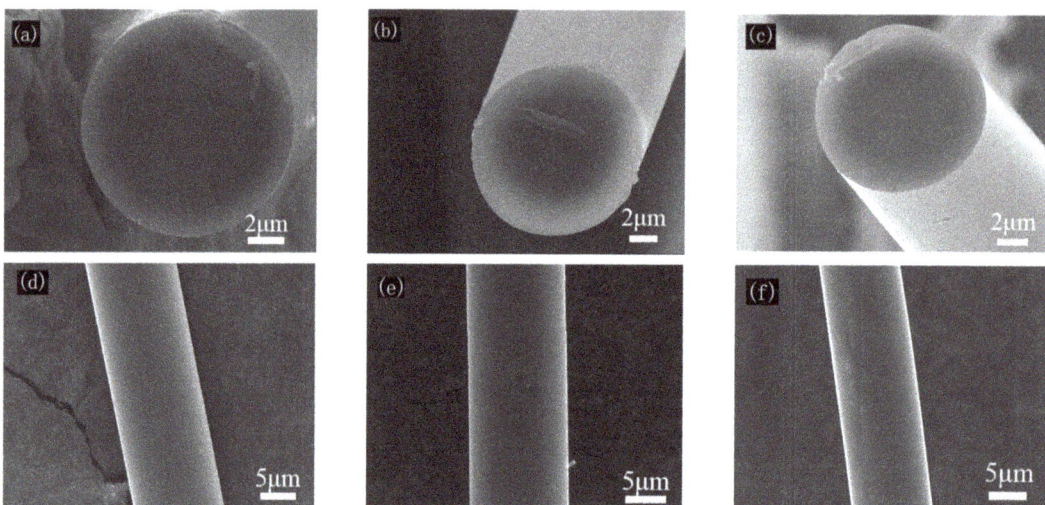

Figure 6. SEM images of the surface and cross sections of the composite fiber: (**a**,**d**) P1-F; (**b**,**e**) P2-F; (**c**,**f**) P3-F.

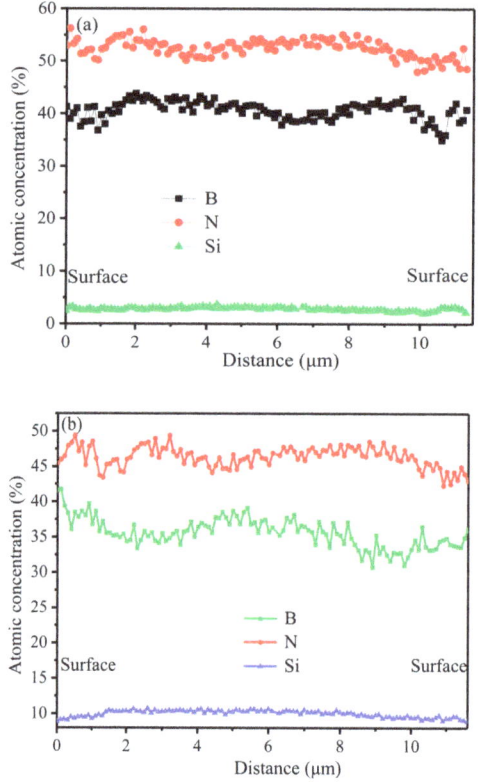

Figure 7. *Cont.*

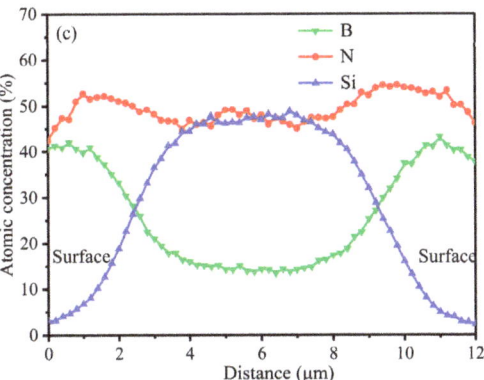

Figure 7. Elemental concentration of B, N and Si along the composite fiber diameter: (**a**) P1-F; (**b**) P2-F; (**c**) P3-F.

The Weibull plots of failure strength of these fibers are illustrated in Figure 8. The tensile strength, Young's modulus, Weibull modulus and dielectric properties (f = 10 GHz) of BN/Si$_3$N$_4$ fibers are listed in Table 4. The tensile strength of BN/Si$_3$N$_4$ fibers rose dramatically with the increase of Si$_3$N$_4$ mass content, and that of P3-F reached 1360 MPa with the Young's modulus of 117 GPa. Compared with the results of Tan et al. [24], the composite fiber (P3-F) showed no h-BN crystals, but a higher tensile strength. Except in the case of the slightly higher mass content of Si$_3$N$_4$ in P3-F, the micro-cracks were caused by the crystallization process of h-BN. Therefore, in order to enhance the tensile strength of BN/Si$_3$N$_4$ composite fibers, the crystallization process of h-BN crystals should be avoided and the mass content of Si$_3$N$_4$ should be enhanced as much as possible in an appropriate ratio range of BN/Si$_3$N$_4$, where Si$_3$N$_4$ could remain amorphous in the final composite fibers. Apart from the excellent tensile strength, the composite fibers (P3-F) showed a low dielectric constant of 3.34 and loss tangent of 0.0047 at 10 GHz. The excellent dielectric properties could be ascribed to the low carbon content [30], which was less than 0.1 wt%. The combination of improved mechanical properties and excellent dielectric behavior demonstrated the potential for wave-transparent applications.

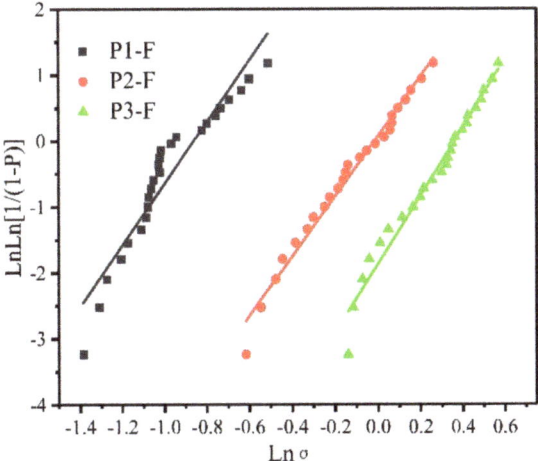

Figure 8. Weibull plot of failure strengths of three composite fiber diameters.

Table 4. Tensile strength, Young's modulus and dielectric properties (f = 10 GHz) of BN/Si_3N_4 composite fibers.

Material	Tensile Strength (MPa)	Young's Modulus (GPa)	Weibull Modulus	Dielectric Constant	Loss Tangent
P1-F	365	35	4.65	3.02	0.0023
P2-F	832	93	4.53	3.21	0.0031
P3-F	1360	117	5.17	3.34	0.0047

4. Conclusions

BN/Si_3N_4 composite ceramics and ceramic fibers were obtained through the precursor-derived ceramic route using the hybrid polymers of poly(tri(methylamino)borazine) (PBN) and polycarbosilane (PCS). When the Si_3N_4 content of the composite ceramics was below 25 wt%, no Si_3N_4 crystals were found at 1600 °C, and the BN/Si_3N_4 composite ceramic containing 25 wt% Si_3N_4 was totally amorphous. Three kinds of BN/Si_3N_4 composite fibers containing 5 wt%, 15 wt% and 25 wt% Si_3N_4 were fabricated successfully, showing the nanocrystallines of BN and amorphous Si_3N_4, of which the mean tensile strength and Young's modulus was enhanced with the increasing of the Si_3N_4 mass content. Additionally, the composite fibers (P3-F) showed a unique BN (shell)/Si_3N_4 (core) structure, with average tensile strength of 1.36 GPa and Young's modulus up to 117 GPa. Moreover, the composite fiber (P3-F) exhibited excellent dielectric properties, with a dielectric constant of 3.34 and a dielectric loss tangent of 0.0047 at 10 GHz. Further research is in progress to optimize the oxygen content to improve the mechanical properties of the composite ceramic fibers.

Author Contributions: Formal analysis, Y.L.; investigation, Y.L. and M.G.; methodology, H.Z. and Z.W.; resources, S.Y., C.H. and W.K.; writing—original draft, Y.L.; writing—review and editing, M.G. and W.Z. All authors have read and agreed to the published version of the manuscript.

Funding: This research was funded by Key Laboratory of Science and Technology on Particle Materials (Grant No. CXJJ-21S043), Key Laboratory of Multiphase Complex Systems (Grant No. MPCS-2021-a-02) and Innovation Academy for Green Manufacture (Grant No. IAGM2020C22), Chinese Academy of Sciences.

Institutional Review Board Statement: Not applicable.

Informed Consent Statement: Not applicable.

Data Availability Statement: Not applicable.

Conflicts of Interest: The authors declare no conflict of interest.

References

1. Tang, L.; Zhang, J.; Tang, Y.; Kong, J.; Liu, T.; Gu, J. Polymer matrix wave-transparent composites: A review. *J. Mater. Sci. Technol.* **2021**, *75*, 225–251. [CrossRef]
2. Tang, L.; Yang, Z.; Tang, Y.; Zhang, J.; Kong, J.; Gu, J. Facile functionalization strategy of PBO fibres for synchronous improving the mechanical and wave-transparent properties of the PBO fibres/cyanate ester laminated composites. *Compos. Part A Appl. Sci. Manuf.* **2021**, *150*, 106622. [CrossRef]
3. Li, B.; Liu, K.; Zhang, C.; Wang, S. Fabrication and properties of borazine derived boron nitride bonded porous silicon aluminum oxynitride wave-transparent composite. *J. Eur. Ceram. Soc.* **2014**, *34*, 3591–3595. [CrossRef]
4. Yang, X.; Li, B.; Li, D.; Shao, C.; Zhang, C. High-temperature properties and interface evolution of silicon nitride fiber reinforced silica matrix wave-transparent composite materials. *J. Eur. Ceram. Soc.* **2019**, *39*, 240–248. [CrossRef]
5. Li, D.; Zhang, C.; Li, B.; Cao, F.; Wang, S.; Li, J. Preparation and properties of unidirectional boron nitride fibre reinforced boron nitride matrix composites via precursor infiltration and pyrolysis route. *J. Mater. Sci. Eng. A* **2011**, *528*, 8169–8173. [CrossRef]
6. Toutois, P.; Miele, P.; Jacques, S.; Cornu, D.; Bernard, S. Structural and mechanical behavior of boron nitride fibers derived from poly[(methylamino)borazine] precursors: Optimization of the curing and pyrolysis procedures. *J. Am. Ceram. Soc.* **2006**, *89*, 42–49. [CrossRef]
7. Bernard, S.; Salameh, C.; Miele, P. Boron nitride ceramics from molecular precursors: Synthesis, properties and applications. *Dalton Trans.* **2016**, *45*, 861–873. [CrossRef] [PubMed]
8. Bernard, S.; Miele, P. Polymer-derived boron nitride: A review on the chemistry, shaping and ceramic conversion of borazine derivatives. *Materials* **2014**, *7*, 7436–7459. [CrossRef] [PubMed]

9. Li, J.; Luo, F.; Zhu, D.; Zhou, W. Influence of Phase Formation on Dielectric Properties of Si_3N_4 Ceramics. *J. Am. Ceram. Soc.* **2007**, *90*, 1950–1952. [CrossRef]
10. Bernard, S.; Miele, P. Nanostructured and architectured boron nitride from boron, nitrogen and hydrogen-containing molecular and polymeric precursors. *Mater. Today* **2014**, *17*, 443–450. [CrossRef]
11. Miele, P.; Bernard, S.; Cornu, D.; Toury, B. Recent developments in polymer-derived ceramic fibers (PDCFs): Preparation, properties and applications—A review. *Soft Mater.* **2007**, *4*, 249–286. [CrossRef]
12. Mou, S.; Liu, Y.; Han, K.; Yu, M. Synthesis and characterization of amorphous SiBNC ceramic fibers. *Ceram. Int.* **2015**, *41*, 11550–11554. [CrossRef]
13. Houska, J. Maximum Achievable N Content in Atom-by-Atom Growth of Amorphous Si-B-C-N Materials. *Materials* **2021**, *14*, 5744. [CrossRef] [PubMed]
14. Han, F.; Wen, H.; Sun, J.; Wang, W.; Fan, Y.; Jia, J.; Chen, W. Tribological Properties of Si_3N_4-hBN Composite Ceramics Bearing on GCr15 under Seawater Lubrication. *Materials* **2020**, *13*, 635. [CrossRef] [PubMed]
15. Lei, Y.; Wang, Y.; Song, Y.; Deng, C. Novel processable precursor for BN by the polymer-derived ceramics route. *Ceram. Int.* **2011**, *37*, 3005–3009. [CrossRef]
16. Al-Ghalith, J.; Dasmahapatra, A.; Kroll, P.; Meletis, E.; Dumitrică, T. Compositional and Structural Atomistic Study of Amorphous Si-B-N Networks of Interest for High-Performance Coatings. *J. Phys. Chem. C* **2016**, *120*, 24346–24353. [CrossRef]
17. Viard, A.; Fonblanc, D.; Schmidt, M.; Lale, A.; Salameh, C.; Soleilhavoup, A.; Wynn, M.; Champagne, P.; Cerneaux, S.; Babonneau, F.; et al. Molecular chemistry and engineering of boron-modified polyorganosilazanes as new processable and functional SiBCN precursors. *Chem. Eur. J.* **2017**, *23*, 9076–9090. [CrossRef] [PubMed]
18. Viard, A.; Fonblanc, D.; Lopez-Ferber, D.; Schmidt, M.; Lale, A.; Durif, C.; Balestrat, M.; Rossignol, F.; Weinmann, M.; Riedel, R.; et al. Polymer derived Si-B-C-N ceramics: 30 years of research. *Adv. Eng. Mater.* **2018**, *20*, 1800360. [CrossRef]
19. Toury, B.; Miele, P.; Cornu, D.; Vincent, H.; Bouix, J. Boron nitride fibers prepared from symmetric and asymmetric alkylaminoborazine. *Adv. Funct. Mater.* **2002**, *12*, 228–234. [CrossRef]
20. Colombo, P.; Mera, G.; Riedel, R.; Soraru, G. Polymer-Derived Ceramics: 40 Years of Research and Innovation in Advanced Ceramics. *J. Am. Ceram. Soc.* **2010**, *93*, 1805–1837. [CrossRef]
21. Tang, Y.; Wang, J.; Li, X.; Xie, Z.; Wang, H.; Li, W.; Wang, X. Polymer-Derived SiBN Fiber for High-Temperature Structural/Functional Applications. *Chem.—Eur. J.* **2010**, *16*, 6458–6462. [CrossRef]
22. Liu, Y.; Peng, S.; Cui, Y.; Chang, X.; Zhang, C.; Huang, X.; Han, K.; Yu, M. Fabrication and properties of precursor-derived SiBN ternary ceramic fibers. *Mater. Des.* **2017**, *128*, 150–156. [CrossRef]
23. Peng, Y.; Han, K.; Zhao, X.; Yu, M. Large-scale preparation of SiBN ceramic fibres from a single source precursor. *Ceram. Int.* **2014**, *40*, 4797–4804. [CrossRef]
24. Tan, J.; Ge, M.; Yu, S.; Lu, Z.; Zhang, W. Microstructure and properties of ceramic fibers of h-BN containing amorphous Si_3N_4. *Materials* **2019**, *12*, 3812. [CrossRef] [PubMed]
25. Zhou, W.; Wang, C.; Ai, T.; Wu, K.; Zhao, F.; Gu, H. A novel fiber-reinforced polyethylene composite with added silicon nitride particles for enhanced thermal conductivity. *Compos. Part A Appl. Sci. Manuf.* **2009**, *40*, 830–836. [CrossRef]
26. Yajima, S.; Hasegawa, Y.; Hayashi, J.; Iimura, M. Synthesis of continuous silicon carbide fibre with high tensile strength and high Young's modulus. *J. Mater. Sci.* **1978**, *13*, 2569–2576.
27. Chen, M.; Ge, M.; Zhang, W. Preparation and properties of hollow BN fibers derived from polymeric precursors. *J. Eur. Ceram. Soc.* **2012**, *32*, 3521–3529. [CrossRef]
28. Yang, H.; Fang, H.; Yu, H.; Chen, Y.; Wang, L.; Jiang, W.; Wu, Y.; Li, J. Low temperature self-densification of high strength bulk hexagonal boron nitride. *Nat. Commun.* **2019**, *10*, 854. [CrossRef]
29. Bernard, S.; Chassagneux, F.; Berthet, M.-P.; Vincent, H.; Bouix, J. Structural and mechanical properties of a high-performance BN fibre. *J. Eur. Ceram. Soc.* **2002**, *22*, 2047–2059. [CrossRef]
30. Román, R.; Hernández, M.; Ibarra, A.; Vila, R.; Mollá, J.; Martín, P.; González, M. The effect of carbon additives on the dielectric behaviour of alumina ceramics. *J. Acta Mater.* **2006**, *54*, 2777–2782. [CrossRef]

Article

High Humidity Response of Sol–Gel-Synthesized BiFeO$_3$ Ferroelectric Film

Yaming Zhang [1], Bingbing Li [2] and Yanmin Jia [1,*]

1 School of Science, Xi'an University of Posts and Telecommunications, Xi'an 710048, China; zymon@163.com
2 School of Communication and Information Engineering, Xi'an University of Posts and Telecommunications, Xi'an 710048, China; lbbing33@163.com
* Correspondence: jiayanmin@xupt.edu.cn; Tel.: +86-8816-6335

Abstract: In this work, a BiFeO$_3$ film is prepared via a facile sol–gel method, and the effects of the relative humidity (RH) on the BiFeO$_3$ film in terms of capacitance, impedance and current–voltage (*I*–*V*) are explored. The capacitance of the BiFeO$_3$ film increased from 25 to 1410 pF with the increase of RH from 30% to 90%. In particular, the impedance varied by more than two orders of magnitude as RH varied between 30% and 90% at 10 Hz, indicating a good hysteresis and response time. The mechanism underlying humidity sensitivity was analyzed by complex impedance spectroscopy. The adsorption of water molecules played key roles at low and high humidity, extending the potential application of ferroelectric BiFeO$_3$ films in humidity-sensitive devices.

Keywords: ferroelectrics; BiFeO$_3$; humidity response; sol–gel preparation

1. Introduction

Humid environments are essential in many fields, such as weather forecasting, agricultural production and personnel health [1,2]. In addition, trace amounts of water molecules can have a significant impact on industry and manufacturing [3,4]. Therefore, it is necessary to explore highly efficient and accurate humidity sensors. Humidity is a physical quantity that indicates the molecular content of water in the air, and is mainly measured by relative humidity. Over the past decade, many techniques for measuring humidity have been reported, including wet and dry bulb hygrometers, piezoelectric quartz films, resistive sensors, and sensors based on current, impedance and surface acoustic waves [2,5]. Among them, impedance-based humidity-sensing technology is the most convenient and commonly used [2]. Impedance humidity sensors work on the principle that changes in humidity can be reflected by changes in the impedance of a hygroscopic medium [6]. Impedance-type humidity sensors have been extensively reported in recent years due to their low cost, fast response speed and small size [6,7]. Impedance measurements indicate that suitable humidity-sensitive materials mainly include polymers, carbon materials and ceramic materials [8,9]. However, polymer films are not suitable for application at high temperatures. Ceramic films with good stability at high temperatures are considered to be the preferred materials for impedance-based humidity sensors due to their unique structure of grain boundaries, grains and pores [10].

For the convenience of microelectronics integration, film materials are often prepared for humidity sensors. Some ferroelectric perovskite (ABO$_3$, where A is a rare earth, alkali or alkaline earth metal and B is a transition metal) humidity-sensing film materials, including BaTiO$_3$, K$_{0.5}$Bi$_{0.5}$TiO$_3$, K$_{0.5}$Na$_{0.5}$NbO$_3$ and LaFeO$_3$, can behave with remarkable humidity-sensing properties [11–13]. BiFeO$_3$ is a well-known lead-free ferroelectric material that has been regarded as a promising spintronic and information-storage receptor material in recent years due to its large remanent polarization and high Curie point [14–16]. BiFeO$_3$ is a distorted perovskite ferroelectric material with a non-stoichiometric ratio, which makes it behave with *p*-type semiconductor behavior and makes it a promising material for high-performance humidity-sensing applications [17,18]. So far, there are few reports exploring the humidity-sensing behavior of $BiFeO_3$ films. In humidity sensors, morphology and cation distribution can be controlled by the synthesis method, which affects the surface reaction. The sol–gel technique is a simple, low-cost and promising method for the preparation of $BiFeO_3$ films [19].

In this work, the capacitance of $BiFeO_3$ film synthesized via the sol–gel method was found to increase from 25 to 1410 pF when RH increased from 30% to 90%. In particular, the impedance varied by more than two orders of magnitude when RH varied between 30% and 90% at 10 Hz, which extends the potential application of ferroelectric $BiFeO_3$ films to humidity-sensitive devices.

2. Materials and Methods

The sol–gel method was used to successfully prepare a $BiFeO_3$ film. Powders of bismuth nitrate ($Bi(NO_3)_3 \cdot 5H_2O$) and ferric nitrate ($Fe(NO_3)_3 \cdot 9H_2O$) were dissolved in $C_3H_8O_2$ solution with a molar ratio of 1:1 and agitated at room temperature for 30 min. Afterwards, enough CH_3COOH was added to the solution for dehydration. During continuous stirring, 5 mL of aminoethanol was added to $BiFeO_3$ solution in order to control the viscosity. Finally, a 0.3 mol/L red-brown mixed solution with a volume of 30 mL was obtained. The mixture was stirred on a magnetic stirrer for 2 h and left at room temperature for 12 h. The obtained reddish-brown $BiFeO_3$ solution was spin-coated on a Pt/Si(111) substrate and dried for 3 min at 180 °C. Then, films were calcined for 20 min at 490 °C. Finally, conductive silver glue was used to stick electrodes on the surface of the $BiFeO_3$ film for the electrical measurement.

The simple structure was determined via XRD (D/Max2550VB+/PC, Japan). The microstructure was characterized via SEM (Nova NanoSEM 450, Lincoln, NE, USA). A ferroelectric analyzer was used to explore the ferroelectric hysteresis loop (Precision Multiferroic, Radiant Technology, Albuquerque, NM, USA). The capacitance and impedance were measured using a precision impedance analyzer (Novocontrol GmbH, Montabaur, Germany). The current–voltage relationship was measured using a current–voltage meter (Agilent B2902A, Santa Clara, CA, USA). A humidifier in an enclosed space was employed to generate an environment with 30% to 90% relative humidity. The RH was measured using a hygrometer.

3. Results and Discussion

3.1. Structure and Morphology of Material

Figure 1 shows the X-ray diffractometer (XRD) patterns of the $BiFeO_3$ film. The diffraction peaks of the pure $BiFeO_3$ sample are consistent with the standard chart of $BiFeO_3$ with rhombohedral *R*3c structure (JCPDS PDF # 86-1518), as shown in Figure 1. There is no impurity peak, which proves that the sample is a pure-phase perovskite structure $BiFeO_3$. The scanning electron microscope image of the $BiFeO_3$ film is shown in the inset of Figure 1. The image reveals that the as-synthesized $BiFeO_3$ film had a porous structure, indicating its excellent ability to adsorb water molecules, which is essential for humidity sensing.

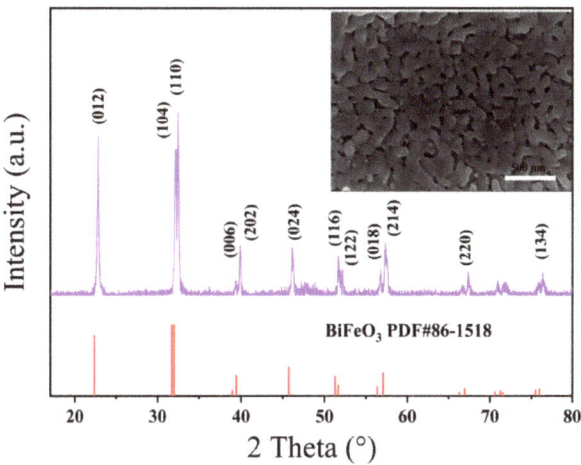

Figure 1. XRD pattern of BiFeO$_3$ film. Inset: SEM image of BiFeO$_3$ film.

The ferroelectric hysteresis loop of the BiFeO$_3$ film is shown in Figure 2. A schematic diagram of the ferroelectric test circuit is shown in the inset of Figure 2. This test circuit was composed of two silver electrode points coated on the surface of the material to connect the wires. In Figure 2, the unsaturated ferroelectric hysteresis loop was obtained due to the serious leakage current [19]. The composition of the sample indicates that the BiFeO$_3$ film had a serious electrical leakage problem due to the multiple valence states of Fe [19].

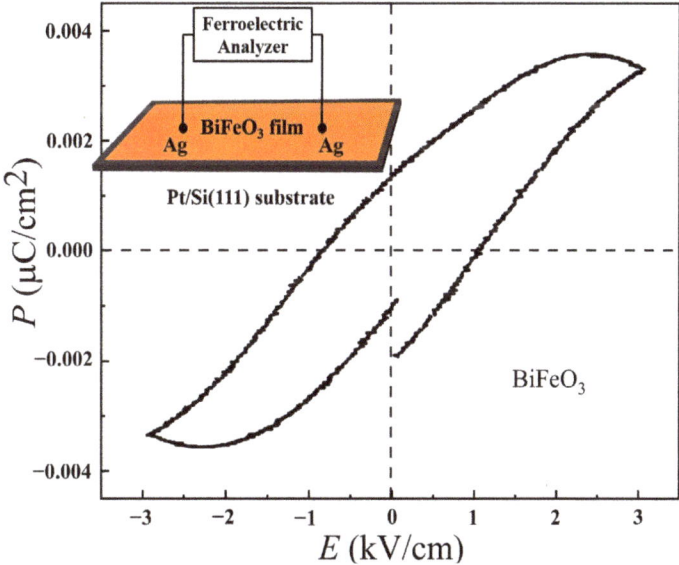

Figure 2. Ferroelectric hysteresis loop. Inset: schematic diagram of the test circuit.

3.2. Humidity-Sensing Properties

The dependence of the capacitance of the BiFeO$_3$ film on the RH was measured at the frequencies of 10, 40, 100, 300, 600 and 1200 Hz, as shown in Figure 3. The inset is a partial enlarged view of capacitance change with RH (RH 30−50%) at different frequencies. At low frequency (i.e., 10 Hz, 40 Hz, 100 Hz), the capacitance increased significantly with increasing

RH. In particular, the capacitance of the BiFeO$_3$ film increased from 25 to 1410 pF as the RH increased from 30% to 90% at 10 Hz. This was due to the increase of physisorbed water molecules on the BiFeO$_3$ film surface with the increase of RH, which made more water molecules polarized. At high frequency (i.e., 300 Hz, 600 Hz, 1200 Hz), the capacitance remained almost constant with increasing RH, implying that frequency is a crucial factor in the humidity response. At high frequency, the dipoles of the water molecules slow their reorientation. The dipole rotation of water molecules no longer resonates with the external field at high frequencies, which means that the polarizability of the water molecules lags behind the frequency of the change of the external electric field. Therefore, the capacitance of the BiFeO$_3$ film had a high humidity response at frequencies range of 10–100 Hz, while RH is independent of the capacitance at frequencies in the range of 100 Hz to 1.2 kHz. The effect of RH on capacitance can be expressed by Equation (1) [20]

$$C = (\varepsilon_\gamma - i \times \frac{\gamma}{\omega \times \varepsilon_0}) \times C_0 \qquad (1)$$

where ε_γ and γ are the permittivity and the electrical conductivity of the BiFeO$_3$ film, respectively. C_0 and ε_0 denote the capacitance of an ideal capacitor and the vacuum permittivity, respectively. C and ω are the capacitance and the frequency, respectively. Equation (1) indicates that the capacitance of the BiFeO$_3$ film is inversely related to ω and is proportional to the material's γ. Both γ and C increase as RH increases [20].

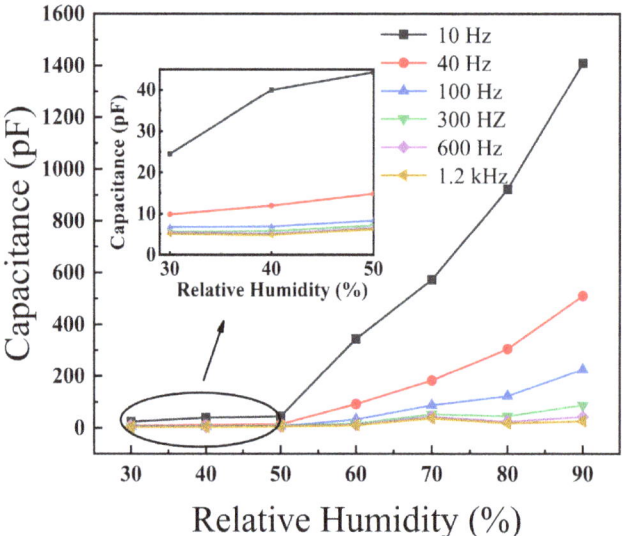

Figure 3. RH dependence on the capacitance of BiFeO$_3$ film. Inset: enlarged capacitance vs. %RH plot (30–50% RH range).

In order to determine the optimal working frequency, the dependence of impedance on RH was measured using BiFeO$_3$ film at 30–90% RH and frequencies of 10, 40, 100, 300, 600 and 1200 Hz, as shown in Figure 4. Since it is difficult to lead the adsorbed water molecules to modify the associated polarization at high frequencies, there was a weak response to humidity at these frequencies. Therefore, it is important to determine the optimal frequency for RH measurements [21]. Figure 4 shows that the impedance of the BiFeO$_3$ film decreased from 1.7×10^5 to 1570 kΩ when RH increased from 30% to 90%. The impedance decreased significantly at 10 Hz, indicating that the optimum working frequency is 10 Hz. Over the entire frequency range, the impedance decreased with the increase of RH. At the same frequency, the impedance change was not obvious at low RH, while the impedance drop was more significant at high RH. This is because the main conduction mechanism for humidity sensing is caused by proton hopping between the sensitive layer of the film and water molecules. At low RH, a small amount of water molecules are chemisorbed on the cations (Bi^{3+} and Fe^{3+}) on the film surface [22]. Due to the lack of a complete adsorption layer, the low polarizability of water molecules eventually leads to high impedance. At high RH, multiple layers of physical adsorption are formed on the basis of the chemical adsorption layer, resulting in the movement of more protons in the water layer [22]. This results in a significant increase in the conductivity of the humidity sensor and a decrease in impedance.

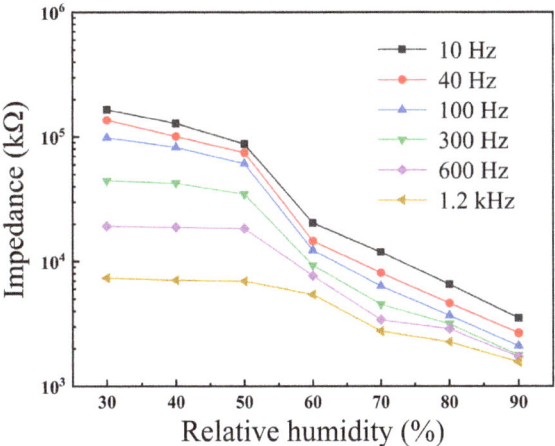

Figure 4. RH dependence on the impedance of $BiFeO_3$ film.

Humidity hysteresis of the $BiFeO_3$ film usually occurred during the desorption of samples. The humidity hysteresis is a critical characteristic for the application of humidity sensing, and is defined as the maximum difference between adsorption and desorption of the humidity sensor. The humidity hysteresis (γH) is expressed in Equation (2) as [21]:

$$\gamma H = \pm \frac{\Delta RH_{MAX}}{2F_{FS}} \quad (2)$$

where RH_{MAX} is the maximum difference in the output of adsorption and desorption processes. F_{FS} is the impedance change over the entire humidity range. The humidity hysteresis characteristics of the $BiFeO_3$ humidity sensor at 10 Hz are shown in Figure 5. It can be seen from the figure that the $BiFeO_3$ showed a narrow hysteresis loop. The $BiFeO_3$ film had a small hysteresis during the entire humidity test with a maximum hysteresis of approximately 16%, mainly caused by residual moisture in the $BiFeO_3$ film layer. With the decrease of RH, the number of water molecules between the layers of the $BiFeO_3$ film gradually decreased, resulting in the gradual disappearance of the hysteresis phenomenon [23,24].

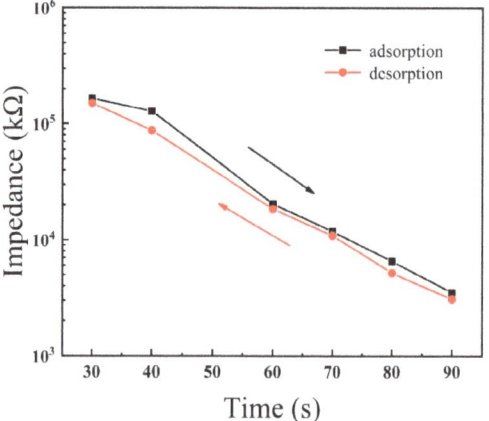

Figure 5. Humidity hysteresis characteristics of BiFeO$_3$ film measured at 10 Hz.

Based on the conversion circuit of a humidity sensor, RH changes in the environment can be converted into an electrical signal that is easy to control and identify. The ideal humidity sensor needs to meet the following characteristics: fast response speed, strong recovery ability and small humidity hysteresis error. The response and recovery times are the times required for the BiFeO$_3$ film to reach 90% of the total impedance change during adsorption and desorption, respectively. Figure 6 shows that the humidity response and recovery times of the BiFeO$_3$ film in the maximum humidity range (30–90% RH) were 60 s and 70 s at 10 Hz, respectively. The recovery time of the BiFeO$_3$ film was higher than the response time due to the higher bonding energy between the adsorbed water molecules and the surface of the sensor material [25]. This result indicates that the BiFeO$_3$ film could rapidly adsorb and desorb water molecules, indicating its potential value for practical applications.

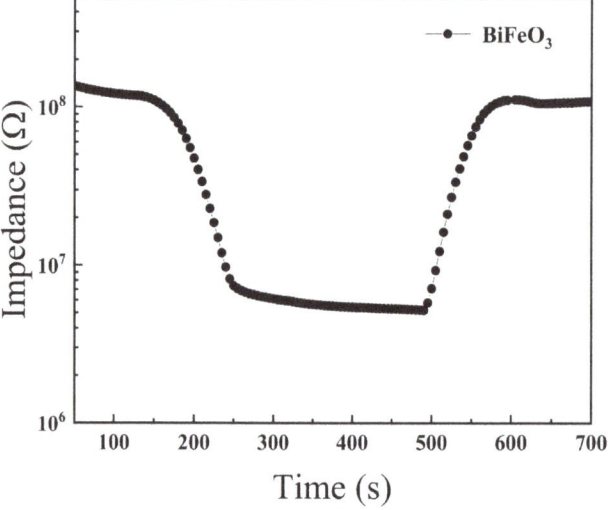

Figure 6. Humidity response and recovery curve of BiFeO$_3$ film measured at 10 Hz.

3.3. Humidity-Sensing Mechanism

The complex impedance curve is an effective method to study the properties of humidity sensing [26]. In AC complex impedance analysis, an AC sinusoidal test signal is applied to a thin-film device, and the frequency of the test signal is changed within a certain range. Figure 7 shows the complex impedance spectrum of the $BiFeO_3$ film in the range of 30–90% RH and in the scanning frequency range of 10–1000 kHz. The complex impedance spectrum of the $BiFeO_3$ film presented a circular arc shape when the RH was lower than 50%, as shown in Figure 7a–c. The complex impedance spectrum gradually changed from a circular arc to a semicircular shape with increasing humidity. Compared to the complex impedance spectra of standard circuit components, it can be concluded that the equivalent circuit diagram for $BiFeO_3$ films in the low-humidity range is composed of parallel connections of resistors and capacitors, as shown in Figure 7h. Oxygen ions and metal ions are exposed on the surface of the $BiFeO_3$ film, and the H_2O molecules on the surface dissociate into H^+ and OH^-. Then, OH^- and H^+ are chemically combined with metal ions and oxygen ions, respectively, to form hydroxyl groups that constitute the first layer of physical adsorption [23]. The charge transfer is carried out according to the Grotthuss chain reaction of $2 H_2O \rightarrow H_3O^+ + OH^-$, which has a weak influence on the capacitance of the $BiFeO_3$ film. H_3O^+ spontaneously transfers H^+ to the second water molecule according to $H_3O^+ \rightarrow H_2O + H^+$ [27,28], and the main mechanism underlying the humidity response is based on proton transport [29].

When RH increased to 70%, the complex impedance spectrum of the $BiFeO_3$ film showed a straight line with a slope of approximately 1 at frequencies from 10 to 100 Hz, as shown in Figure 7d,e. On top of the first layer of physical adsorption, more adsorption layers are formed through hydrogen bonding to generate a liquid water layer, and the physical adsorption changes from single-layer to multi-layer [30]. When RH increased to 90%, the proportion of the straight-line part of the complex impedance spectrum increased, while the semicircle part was compressed. The appearance of a straight line in the low-frequency region of the complex impedance spectrum indicates that the $BiFeO_3$ film has a significant Warburg impedance due to ion diffusion, as shown in Figure 7f,g. The corresponding equivalent circuit includes resistance, capacitance and Warburg impedance, as shown in Figure 7i. With the continuous increase in the number of adsorbed water molecules, the adsorption on the sample surface evolves into multi-molecular layer adsorption. The surface of the $BiFeO_3$ films is covered by water, resulting in a rapid increase in the amount of H^+, which further increases the conductivity [31,32].

The *I–V* characteristics of the $BiFeO_3$ film at different RH levels are presented in Figure 8. The inset is a partial enlarged view of the change in *I–V* with RH. At different RHs, $BiFeO_3$ film exhibited linear *I–V* characteristics, which indicates an ohmic contact between the $BiFeO_3$ film surface and electrodes. Since the resistance was constant over the range of supply voltage, the sensitivity was the same regardless of the operating bias, which allows operation at low power in practical application [33]. As RH increased, the conductivity of the $BiFeO_3$ film increased, resulting in a decrease in current. The excellent humidity response makes $BiFeO_3$ films a potential candidate for practical humidity-sensing applications.

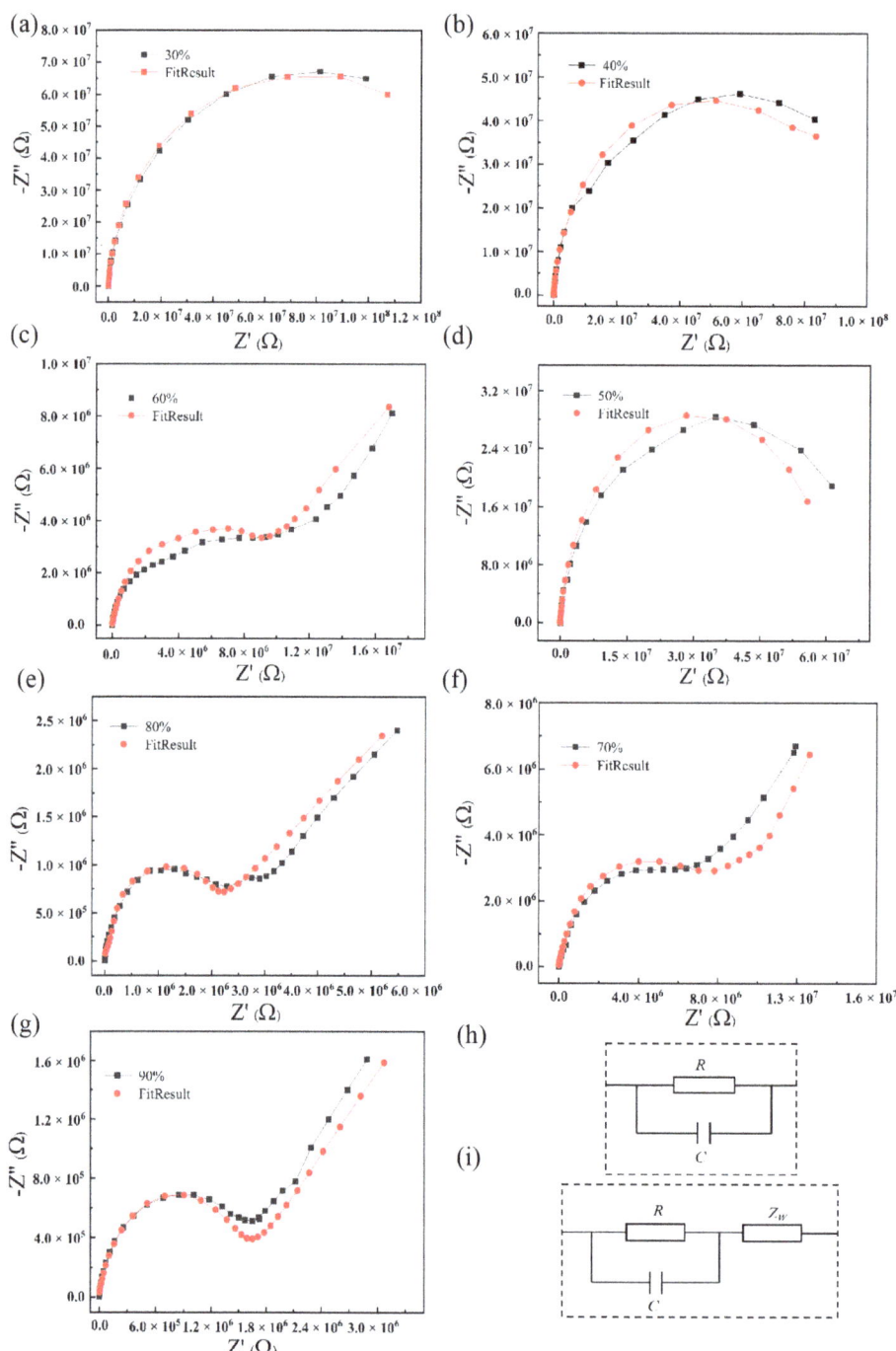

Figure 7. The complex impedance properties of BiFeO$_3$ film under different humidities. (**a**) 30% RH; (**b**) 40% RH; (**c**) 50% RH; (**d**) 60% RH; (**e**) 70% RH; (**f**) 80% RH; (**g**) 90% RH. (**h**) The equivalent circuit fit by Zview in the 30–50% RH range. (**i**) The equivalent circuit fit by Zview in the 60–90% RH range.

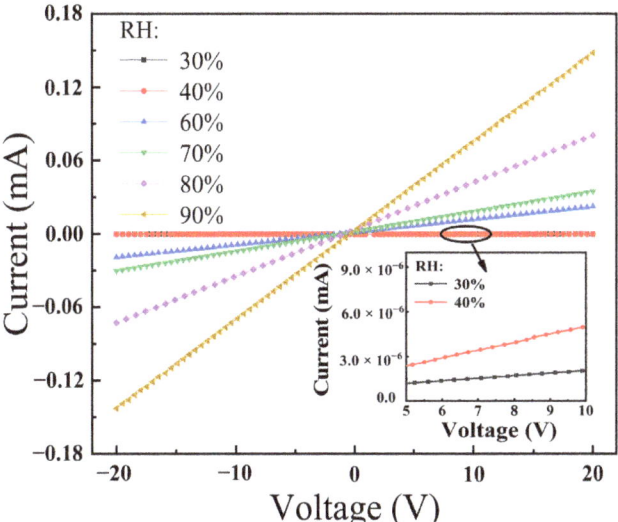

Figure 8. Dependence of current on voltage for the BiFeO$_3$ film at various RHs. Inset: the enlarged I–V vs. %RH plot (30–40% RH range).

4. Conclusions

The BiFeO$_3$ film prepared in this study via a simple sol–gel method exhibited significant humidity sensitivity with capacitance and impedance changes of nearly 2–3 orders of magnitude as RH increased from 30% to 90%. In the whole humidity range, the experimental results of humidity hysteresis and humidity response recovery indicate that BiFeO$_3$ film is an excellent material for application in humidity sensors.

Author Contributions: B.L. conceived and designed the experiments; Y.Z. and Y.J. revised the paper and contributed materials/reagents. All authors have read and agreed to the published version of the manuscript.

Funding: This work was supported by the National Natural Science Foundation of China (Grant Numbers: 51872264, 22179108), Shaanxi Provincial Natural Science Foundation of China (Grant Number: 2020JM-579), Key Research and Development Projects of Shaanxi Province (Grant Number: 2020GXLH-Z-032).

Institutional Review Board Statement: Not applicable.

Informed Consent Statement: Not applicable.

Data Availability Statement: The data presented in this study are available on request from the corresponding author.

Conflicts of Interest: There are no conflict to declare.

References

1. Qi, R.; Lin, X.; Dai, J.; Zhao, H.; Liu, S.; Fei, T.; Zhang, T. Humidity sensors based on MCM-41/polypyrrole hybrid film via in-situ polymerization. *Sens. Actuators B Chem.* **2018**, *277*, 584–590. [CrossRef]
2. Zhang, Y.; Duan, Z.; Zou, H.; Ma, M. Drawn a facile sensor: A fast response humidity sensor based on pencil-trace. *Sens. Actuators B Chem.* **2018**, *261*, 345–353. [CrossRef]
3. Si, R.J.; Li, T.Y.; Sun, J.; Wang, J.; Wang, S.T.; Zhu, G.B.; Wang, C.C. Humidity sensing behavior and its influence on the dielectric properties of (In + Nb) co-doped TiO$_2$ ceramics. *J. Mater. Sci.* **2019**, *54*, 14645–14653. [CrossRef]
4. Wang, J.; Guo, Y.M.; Wang, S.T.; Tong, L.; Sun, J.; Zhu, G.B.; Wang, C.C. The effect of humidity on the dielectric properties of (In+Nb) co-doped SnO$_2$ ceramics. *J. Eur. Ceram. Soc.* **2019**, *39*, 323–329. [CrossRef]
5. Duan, Z.; Zhao, Q.; Wang, S.; Yuan, Z.; Zhang, Y.; Li, X.; Wu, Y.; Jiang, Y.; Tai, H. Novel application of attapulgite on high performance and low-cost humidity sensors. *Sens. Actuators B Chem.* **2020**, *305*, 127534. [CrossRef]

6. Wu, Z.; Yang, J.; Sun, X.; Wu, Y.; Wang, L.; Meng, G.; Kuang, D.; Guo, X.Z.; Qu, W.; Du, B.; et al. An excellent impedance-type humidity sensor based on halide perovskite CsPbBr$_3$ nanoparticles for human respiration monitoring. *Sens. Actuators B Chem.* **2021**, *337*, 129772. [CrossRef]
7. Weng, Z.; Qin, J.; Umar, A.A.; Wang, J.; Zhang, X.; Wang, H.; Cui, X.; Li, X.; Zheng, L.; Zhan, Y. Lead-free Cs$_2$BiAgBr$_6$ double perovskite-based humidity sensor with superfast recovery time. *Adv. Funct. Mater.* **2019**, *29*, 1902234. [CrossRef]
8. Cho, M.-Y.; Kim, S.; Kim, I.-S.; Kim, E.-S.; Wang, Z.-J.; Kim, N.-Y.; Kim, S.-W.; Oh, J.-M. Perovskite-induced ultrasensitive and highly stable humidity sensor systems prepared by aerosol deposition at room temperature. *Adv. Funct. Mater.* **2020**, *30*, 1907449. [CrossRef]
9. Dai, J.; Zhang, T.; Zhao, H.; Fei, T. Preparation of organic-inorganic hybrid polymers and their humidity sensing properties. *Sens. Actuators B Chem.* **2017**, *242*, 1108–1114. [CrossRef]
10. Farahani, H.; Wagiran, R.; Urban, G.A. Investigation of room temperature protonic conduction of perovskite humidity sensors. *IEEE Sens. J.* **2020**, *21*, 9657–9666. [CrossRef]
11. Kumar, A.; Wang, C.; Meng, F.-Y.; Liang, J.-G.; Xie, B.-F.; Zhou, Z.-L.; Zhao, Z.; Kim, N.-Y. Aerosol deposited BaTiO$_3$ film based interdigital capacitor and squared spiral capacitor for humidity sensing application. *Ceram. Int.* **2021**, *47*, 510–520. [CrossRef]
12. Zhao, J.; Liu, Y.; Li, X.; Lu, G.; You, L.; Liang, X.; Liu, F.; Zhang, T.; Du, Y. Highly sensitive humidity sensor based on high surface area mesoporous LaFeO$_3$ prepared by a nanocasting route. *Sens. Actuators B Chem.* **2013**, *181*, 802–809. [CrossRef]
13. Wang, N.; Luo, X.; Han, L.; Zhang, Z.; Zhang, R.; Olin, H.; Yang, Y. Structure, performance, and application of BiFeO$_3$ nanomaterials. *Nano-Micro Lett.* **2020**, *12*, 81. [CrossRef] [PubMed]
14. Yang, S.Y.; Zhang, F.; Xie, X.; Sun, H.; Zhang, L.; Fan, S. Enhanced leakage and ferroelectric properties of Zn-doped BiFeO$_3$ thin films grown by sol-gel method. *J. Alloy Compd.* **2018**, *734*, 243–249. [CrossRef]
15. Chen, M.; Jia, Y.; Li, H.; Wu, Z.; Huang, T.; Zhang, H. Enhanced pyrocatalysis of the pyroelectric BiFeO$_3$/g-C$_3$N$_4$ heterostructure for dye decomposition driven by cold-hot temperature alternation. *J. Adv. Ceram* **2021**, *10*, 338–346. [CrossRef]
16. Wu, J.; Mao, W.; Wu, Z.; Xu, X.; You, H.; Xue, A.; Jia, Y.A.X. Strong pyro-catalysis of pyroelectric BiFeO$_3$ nanoparticles under a room-temperature cold-hot alternation. *Nanoscale* **2016**, *8*, 7343–7350. [CrossRef]
17. You, H.; Jia, Y.; Wu, Z.; Xu, X.; Qian, W.; Xia, Y.; Ismail, M. Strong piezo-electrochemical effect of multiferroic BiFeO$_3$ square micro-sheets for mechanocatalysis. *Electrochem. Commun.* **2017**, *79*, 55–58. [CrossRef]
18. Preethi, A.J.; Ragam, M. Effect of doping in multiferroic BFO: A review. *J. Adv. Dielect.* **2021**, *11*, 2130001. [CrossRef]
19. Xu, X.; Xiao, L.; Haugen, N.O.; Wu, Z.; Jia, Y.; Zhong, W.; Zou, J. High humidity response property of sol-gel synthesized ZnFe$_2$O$_4$ films. *Mater. Lett.* **2018**, *213*, 266–268. [CrossRef]
20. Liang, J.-G.; Kim, E.-S.; Wang, C.; Cho, M.-Y.; Oh, J.-M.; Kim, N.-Y. Thickness effects of aerosol deposited hygroscopic films on ultra-sensitive humidity sensors. *Sens. Actuators B Chem.* **2018**, *265*, 632–643. [CrossRef]
21. Gong, M.; Li, Y.; Guo, Y.; Lv, X.; Dou, X. 2D TiO$_2$ nanosheets for ultrasensitive humidity sensing application benefited by abundant surface oxygen vacancy defects. *Sens. Actuators B Chem.* **2018**, *262*, 350–358. [CrossRef]
22. Rachida, D.; Nouara, L.; M'hand, O.; Malika, S.; Yannick, G.; Ahcène, C.; Bertrand, B. Improvement of humidity sensing performance of BiFeO$_3$ nanoparticles-based sensor by the addition of carbon fibers. *Sens. Actuator A Phys.* **2020**, *307*, 111981.
23. Mahapatra, P.L.; Das, S.; Mondal, P.P.; Das, T.; Saha, D.; Pal, M. Microporous copper chromite thick film based novel and ultrasensitive capacitive humidity sensor. *J. Alloy Compd.* **2021**, *859*, 157778. [CrossRef]
24. Duan, Z.; Zhao, Q.; Wang, S.; Huang, Q.; Yuan, Z.; Zhang, Y.; Jiang, Y.; Tai, H. Halloysite nanotubes: Natural, environmental-friendly and low-cost nanomaterials for high-performance humidity sensor. *Sens. Actuators B Chem.* **2020**, *317*, 128204. [CrossRef]
25. Liang, J.-G.; Wang, C.; Yao, Z.; Liu, M.-Q.; Kim, H.-K.; Oh, J.-M.; Kim, N.-Y. Preparation of ultrasensitive humidity-sensing films by aerosol deposition. *ACS Appl. Mater. Interfaces* **2018**, *10*, 851–863. [CrossRef]
26. Zia, T.H.; Ali Ahsh, A.H. Understanding the adsorption of 1 NLB antibody on polyaniline nanotubes as a function of zeta potential and surface charge density for detection of hepatitis C core antigen: A label-free impedimetric immunosensor. *Colloids Surf. A Physicochem. Eng. Asp.* **2021**, *626*, 127076. [CrossRef]
27. Li, T.Y.; Si, R.J.; Sun, J.; Wang, S.T.; Wang, J.; Ahmed, R.; Zhu, G.B.; Wang, C.C. Giant and controllable humidity sensitivity achieved in (Na+ Nb) co-doped rutile TiO$_2$. *Sens. Actuators B Chem.* **2019**, *293*, 151–158. [CrossRef]
28. Si, R.; Xie, X.; Li, T.; Zheng, J.; Cheng, C.; Huang, S.; Wang, C. TiO$_2$/(K, Na)NbO$_3$ nanocomposite for boosting humidity-sensing performances. *ACS Sens.* **2020**, *5*, 1345–1353. [CrossRef]
29. Mallick, S.; Ahmad, Z.; Qadir, K.W.; Rehman, A.; Shakoor, R.A.; Touati, F.; Al-Muhtaseb, S.A. Effect of BaTiO$_3$ on the sensing properties of PVDF composite-based capacitive humidity sensors. *Ceram. Int.* **2020**, *46*, 2949–2953. [CrossRef]
30. Nikolic, M.V.; Krstic, J.B.; Labus, N.J.; Lukovic, M.D.; Dojcinovic, M.P.; Radovanovic, M.; Tadic, N.B. Structural, morphological and textural properties of iron manganite (FeMnO$_3$) thick films applied for humidity sensing. *Mater. Sci. Eng. B* **2020**, *257*, 114547. [CrossRef]
31. Ji, G.-J.; Zhang, L.-X.; Zhu, M.-Y.; Li, S.-M.; Yin, J.; Zhao, L.-X.; Fahlman, B.D.; Bie, L.-J. Molten-salt synthesis of Ba$_{5-x}$Sr$_x$Nb$_4$O$_{15}$ solid solutions and their enhanced humidity sensing properties. *Ceram. Int.* **2018**, *44*, 477–483. [CrossRef]
32. Ma, H.; Fang, H.; Wu, W.; Zheng, C.; Wu, L.; Wang, H. A highly transparent humidity sensor with fast response speed based on α-MoO$_3$ thin films. *RSC Adv.* **2020**, *10*, 25467–25474. [CrossRef]
33. Han, J.-W.; Kim, B.; Li, J.; Meyyappan, M. Carbon nanotube based humidity sensor on cellulose paper. *J. Phys. Chem. C* **2012**, *116*, 22094–22097. [CrossRef]

Article

The Tribological Behaviors in Zr-Based Bulk Metallic Glass with High Heterogeneous Microstructure

Yubai Ma [1,2,*], Mei Li [3] and Fangqiu Zu [2,*]

1. College of Chemistry and Chemical Engineering, Chongqing University, Chongqing 401133, China
2. Liquid/Solid Metal Processing Institute, School of Materials Science and Engineering, Hefei University of Technology, Hefei 230009, China
3. College of Materials Science and Engineering, Chongqing University, Chongqing 401133, China
* Correspondence: mayb168@hotmail.com (Y.M.); fangqiuzu@hotmail.com (F.Z.)

Abstract: Microstructural inhomogeneity of bulk metallic glasses (BMGs) plays a significant role in their mechanical properties. However, there is hardly ant research concerning the influence of heterogeneous microstructures on tribological behaviors. Hence, in this research, the tribological behaviors of different microstructural-heterogeneity BMGs sliding in-air were systematically investigated, and the corresponding wear mechanisms were disclosed via analyzing the chemical composition and morphology of the wear track. Higher microstructural-heterogeneity BMGs can possess a better wear resistance both under dry sliding and a 3.5% NaCl solution. The results suggest that microstructural heterogeneity enhancement is a valid strategy to improve the tribological performance of BMGs.

Keywords: BMGs; microstructural inhomogeneity; tribological behaviors; HRRF

Citation: Ma, Y.; Li, M.; Zu, F. The Tribological Behaviors in Zr-Based Bulk Metallic Glass with High Heterogeneous Microstructure. *Materials* 2022, 15, 7772. https://doi.org/10.3390/ma15217772

Academic Editors: Qingyuan Wang and Yu Chen

Received: 27 September 2022
Accepted: 1 November 2022
Published: 4 November 2022

Publisher's Note: MDPI stays neutral with regard to jurisdictional claims in published maps and institutional affiliations.

Copyright: © 2022 by the authors. Licensee MDPI, Basel, Switzerland. This article is an open access article distributed under the terms and conditions of the Creative Commons Attribution (CC BY) license (https://creativecommons.org/licenses/by/4.0/).

1. Introduction

Bulk metallic glasses (BMGs) exhibit superior mechanical and physical properties because of their disordered atomic microstructure [1–3]. Consequently, many researchers have been attracted to developing BMGs for their potential applications. However, most scholars focus on the plasticity of BMGs. As a potential structural material, the tribological property is a very significant performance factor in engineering equipment with relative motions in service [4], such as artificial bones [5], golf clubs [6], gear wheels, etc. With very high hardness and elasticity, BMGs are considered superior as wear resistant materials [1]. According to current research, some studies demonstrate that BMGs possess a much longer lifetime in wear applications than crystalline materials [7,8]. For example, A. Inoue et al. reported that micro-sized bearing rollers made of Ni-based BMGs exhibited a lifetime of 2500 h compared with 8 h for SK-steel [9]. However, contradictory conclusions appeared in other experiments [10,11]. For instance, Tam et al. found that Cu-based BMG had a worse frictional coefficient and wear rate than AISI 304 stainless steel under dry and 3.5% NaCl solution, even though Cu-based BMG showed a higher hardness [5]. Additionally, further studies on the atomic-scale and nanoscale scratch wear resistance of a $Cu_{47}Zr_{45}Al_8$ bulk metallic glass by S.V. Ketov et al. identified that the wear rate is found to be significantly reduced by the formation of native and artificially grown surface oxides, indicating that surface oxides hold better wear resistance than $Cu_{47}Zr_{45}Al_8$ bulk metallic glass [12]. Furthermore, increasing efforts have been made to study the tribological behaviors of BMG under seawater. Since seawater is generally simulated with a 3.5% NaCl solution, the tribological behaviors of BMG should also be studied in 3.5% NaCl [13]. Hence, it is necessary to explore a way to further strengthen BMGs' wear resistance both under dry sliding and 3.5% NaCl solution.

Nevertheless, contrary to crystalline metallic materials, the brittleness of BMGs could promote crack propagation and aggravate the delamination of the oxide layers during the wear behavior, resulting in a serious weakening of their tribological behaviors [14–16].

From this point of view, overcoming brittleness becomes greatly significant. Previous studies demonstrate that fabricating the high microstructural inhomogeneity in BMGs is an emerging strategy for remarkably increasing plasticity [17–19].

Inspired by this strategy, we have developed the high rheological rate forming method (HRRF), which could improve plasticity of BMGs by modulating the microstructural heterogeneity of BMGs [17]. Among the various BMGs, Zr-based BMGs have been promising as structural components in many fields due to their high glass-forming abilities, high strength, and high elastic strain [4]. Yet, building on this approach, we changed the microstructural heterogeneity of $Zr_{54.46}Al_{9.9}Ni_{4.95}Cu_{29.7}Pd_{0.99}$ BMGs by HRRF to systematically clarify more information between tribological behaviors and the microstructural heterogeneity of BMGs under dry sliding conditions and in 3.5% NaCl solution.

2. Materials and Methods

$Zr_{54.46}Al_{9.9}Ni_{4.95}Cu_{29.7}Pd_{0.99}$ BMG was prepared through arc melting under a high-purity argon atmosphere. To achieve homogeneity of composition, the master alloys were melted at least four times. Then, a BMG rod with a diameter of 6 mm was produced by copper mold suction casting and cut into a length of 27 mm for HRRF. The wear experiment was performed using an MMW-1A pin-on-disk apparatus in open air under dry conditions and with a 3.5% NaCl solution. BMG samples with a size of $\varphi 4 \times 13$ mm were machined as wear pins to be rubbed against a rotating steel disk of micro-hardness (HV) 728. In the pin-on-disk test, the normal load was 40 N while sliding at a speed of 0.13 m/s, and the sliding duration was 1800 s. The surface microstructure of the samples was recorded using a JSM-6490LV scanning electron microscope (SEM) with energy dispersive X-ray spectrometry (EDS) and a Cypher S atomic force microscope (AFM). All the pin surfaces of samples and disk surfaces were polished using diamond paste. Under each given condition, three samples were measured to ensure the reliability of the data. The glass structure of alloys was confirmed by an X-ray diffractometer (XRD) with Cu Kα radiation. After a wear experiment, each specimen was cleaned ultrasonically. The weight change of each sample before and after the friction tests was determined using an AUY 120 balance with a precision of 0.0001 g. The wear rate was calculated as [20].

$$W = \frac{V}{F} \times S \quad (1)$$

where V is the wear volume loss of the sample, F is the applied normal load, and S is the total sliding distance.

The nano-indentations were carried out over a square area of 48×48 μm^2 for the as-cast, and samples were treated with a Berkovich diamond tip. The constant depth was 300 nm. Each nano-indentations were penetrated at the same depth (300 nm) at a constant loading rate of 0.05 s^{-1}, and the spacing between adjacent indentations was 6 μm.

The HRRF was introduced to manipulate the microstructural heterogeneity of BMGs. As reported in our previous study [19], this fabrication process consists of three steps, heating up to the supercooled liquid region through fast Joule heating, squeezing the supercooled liquid into a copper mold cavity under the preset load, and then rapidly cooling it down to room temperature in the copper mold.

3. Results and Discussions

3.1. Microstructure

Figure 1 illustrates the X-ray diffraction pattern of as-cast and treated $Zr_{54.46}Al_{9.9}Ni_{4.95}Cu_{29.7}Pd_{0.99}$ BMGs. It shows a broad halo with the absence of detectable crystalline peaks, indicating that the treated sample remained a glassy structure.

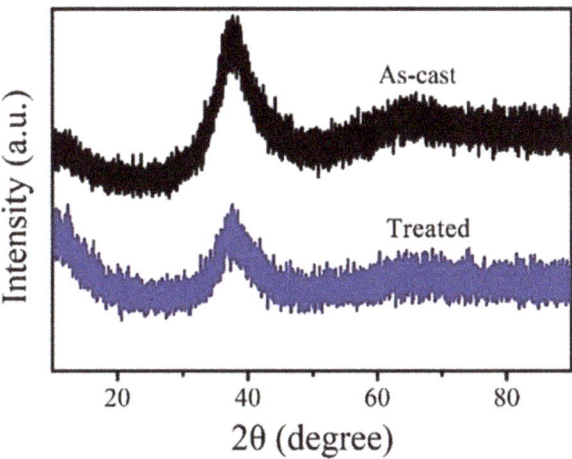

Figure 1. XRD patterns of $Zr_{54.46}Al_{9.9}Ni_{4.95}Cu_{29.7}Pd_{0.99}$ BMGs before and after HRRF treatment [17].

To gain a better understanding of heterogeneous microstructure for as-cast and treated $Zr_{54.46}Al_{9.9}Ni_{4.95}Cu_{29.7}Pd_{0.99}$ BMGs, the distributions of the nano-hardness were obtained, as shown in Figure 2. It is obvious that the as-cast material is rather homogeneous, with hardness values ranging between 7.45 and 7.91 GPa. Conversely, the treated sample is much more heterogeneous and displays a wider range of hardness values (7.288–8.164 GPa). To a certain extent, these distributions demonstrated that the microstructure of treated BMGs can be considered more inhomogeneous than the as-cast samples. Confirmed by early studies, this phenomenon can be interpreted by Lennard-Jones-like potential function. During HRRF-treating, the drastic amount of mechanical work by the preset load was intruded into BMGs driving short-range atomic rearrangement [17,20], leading to hard regions with a higher atomic packing density, and soft regions with a lower atomic packing density.

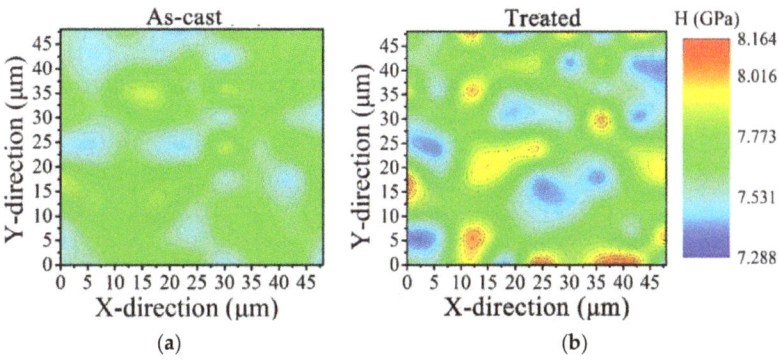

Figure 2. (a,b) are the distribution of the nano-indentation hardness of the as-cast [17] and treated sample, respectively.

3.2. Roughness of Pre-Tested Surfaces

It is well known that roughness of pre-tested surfaces would interfere with the friction properties of BMGs [21]. Therefore, before friction tests, AFM are used to evaluate pre-tested surfaces. The micro-morphology, pre-tested surfaces of the as-cast and treated BMGs are analyzed by AFM as illustrated in Figure 3a,b. It is intuitively found that all pre-tested surfaces exhibit the interlaced topography of "peak" and "valley" on the pre-tested surfaces. Significantly, the relative width of the pre-tested surface roughness for the as-cast glass is 16.6 nm, much the same as that (18.4 nm) in the treated sample. Moreover, by further calculation, the average surface roughness of pre-tested surfaces (Ra) of the cast sample is 4.61 nm, and the Ra of the treated sample is 4.9 nm. Such tiny changes suggest that the pre-tested surfaces exhibit ideal flatness, minimizing the effect of the surface roughness of pre-tested surfaces on the accuracy of friction experiments.

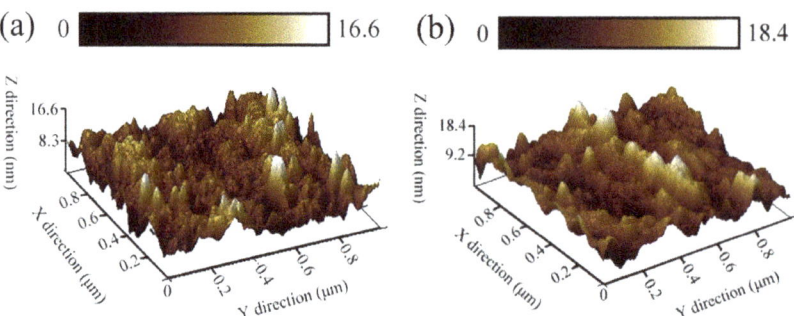

Figure 3. (a,b) are are high-magnification detailed micrographs of pre-tested surfaces for as-cast and treated samples to be tested under AFM.

3.3. Wear Performance

As shown in Figure 2, HRRF could enhance the microstructural heterogeneity of $Zr_{54.46}Al_{9.9}Ni_{4.95}Cu_{29.7}Pd_{0.99}$ BMGs. Thus, the as-cast and treated BMGs were tested for wear resistance. Figure 4 presents the friction coefficient curves of as-cast and treated BMGs under dry conditions and 3.5% NaCl solution. As described in Figure 4a, it was found that both curves have a steady-state stage after an initial rapid increasing period under the dry-friction condition; the coefficients of friction in the steady-state stage are ~0.578 and ~0.676 for as-cast and treated samples respectively. The treated BMG displays a higher friction coefficient (COF). This phenomenon may be interpreted as that during the friction experiment, the harder regions in treated BMGs can remain intact for a long time, which increases the relative movement resistance between the material and the friction pair [22]. However, in Figure 4b, under the 3.5% NaCl solution condition, the friction coefficients of all samples exhibit around 0.29, and much lower, values than those in the air condition, owing to the lubrication effect of the NaCl solution.

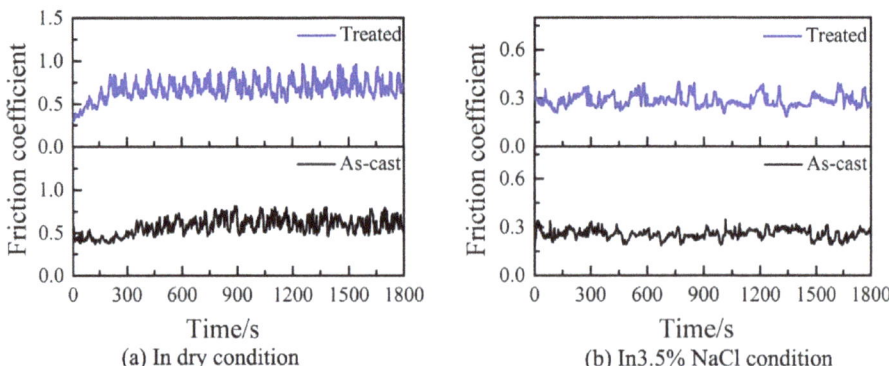

Figure 4. The friction coefficient of as-cast and treated $Zr_{54.46}Al_{9.9}Ni_{4.95}Cu_{29.7}Pd_{0.99}$ BMGs as a function of sliding time under (**a**) dry and (**b**) 3.5% NaCl solution.

The wear rate is always regarded as an important parameter to evaluate the wear resistance of the materials [6,23]. Thus, the wear rate of as-cast and treated BMGs under dry and

3.5% NaCl solution are illustrated in Figure 5. The wear rates of as-cast BMG and treated BMG in dry are 22.1×10^{-6} mm$^3 \cdot$N^{-1}m^{-1} and 14.7×10^{-6} mm$^3 \cdot$N^{-1}m^{-1}, respectively. In addition, the wear rates of BMGs sliding in 3.5% NaCl solution decreased in the following order: 4.5×10^{-6} mm$^3 \cdot$N^{-1}m^{-1} for the as-cast BMG, and 3×10^{-6} mm$^3 \cdot$N^{-1}m^{-1} for the treated BMG. Evidently, the values of wear rates of as-cast samples show remarkably higher values than those of the treated samples under different conditions.

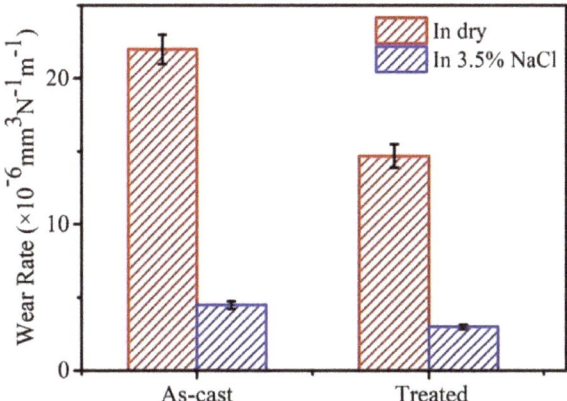

Figure 5. The wear rate of as-cast and treated $Zr_{54.46}Al_{9.9}Ni_{4.95}Cu_{29.7}Pd_{0.99}$ BMGs in different environments.

To be specific, under the dry conditions, due to the direct contact friction between the friction pair of metals, the friction-induced heat easily appears on the tested surfaces of samples [24]. With the heat accumulated constantly, the worn surface not only becomes soft, but also reacts with air to form oxide layers. The hardness of the oxide layers is higher than that of the BMGs, thereby improving the wear resistance of the material. However, due to the high hardness and limited ductility of the oxide layer, it is easy to induce the separation of the oxide layer from the BMGs matrix under the action of sliding shear force, which generates oxide layers which can wear out more easily than BMGs [25,26]. BMGs treated by HRRF would show a supernal microstructural inhomogeneity, which indicates that the hard regions become stronger and the soft regions become more fragile compared with the as-cast BMGs. Hence, for treated BMGs, because of the higher-density arrangement of atoms, it is difficult to oxidize the hard regions. As a result, the number of oxide layers generated on the worn surface of the treated sample may be less than those of the as-cast sample. Eventually, the treated BMGs exhibit higher wear resistance. Similarly, under 3.5% NaCl solution, although there are only a few oxide layers on the worn surfaces of BMGs for a good cooling effect of the solution, due to harder regions, the treated BMGs still show a better wear resistance ability. All results evidence that enhancing the microstructural inhomogeneity of BMGs can improve the wear resistance effectively in these two external environments.

3.4. Worn Surfaces Analysis

To intuitively understand the differences in wear performance between the as-cast and treated BMGs, the morphologies of the wear scars were further examined by the SEM. Figure 6a,c display the worn surface micrographs of as-cast and treated BMGs under dry conditions; the deep and wide grooves with flake-like wear debris parallel to the sliding direction can be clearly observed on the worn surfaces of all BMGs, which suggests that the wear mechanism is mainly controlled by the integration of abrasive wear and adhesive wear [27]. Apparently, at the beginning of the dry-sliding, many wear debris particles appear between the friction pair and the sample. This debris may act as abrasive particles to plow the renewed surface, trap between the contact surfaces, and squeeze into the subsurface, resulting in the formation of ploughed grooves [28]. As the friction test goes on, these wear debris particles can be gradually converged and softened (or liquefied) once the temperature of frictional heating on the wear surface is elevated close to the temperature of glass transition. Eventually, they turn into large, flake-like wear debris which attaches to the wear surface of as-cast and treated samples. Furthermore, upon closer inspection, there are serious delamination, peeling, and micro-cracks on the wear surface of as-cast samples (as illustrated in Figure 6b), indicating that the highly localized stress concentration is induced on this surface on account of the brittleness of as-cast BMGs [29]. However, as shown in Figure 6d, there is no delamination or cracks appearing on the worn surfaces of treated BMGs, due to the higher inhomogeneous structure which would possess the higher ductility. When subjected to compressive stress or shear stress during the process of friction, the treated samples are more likely to undergo plastic deformation, effectively avoiding the appearance of delamination, peeling, and cracks.

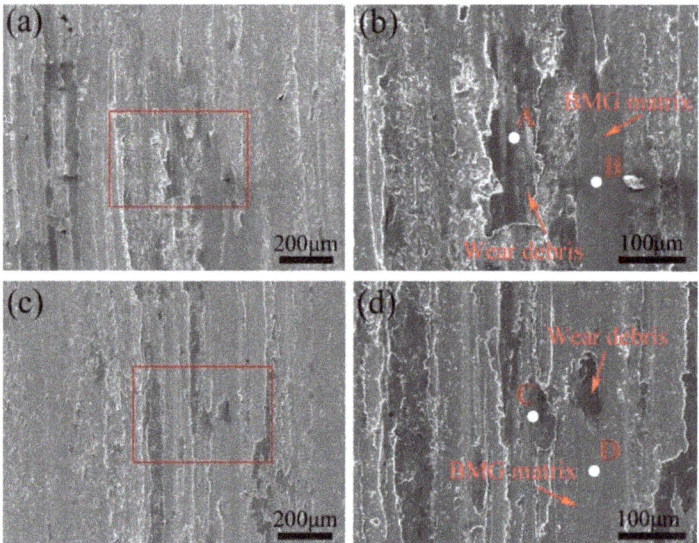

Figure 6. (**a**,**c**) SEM images of worn scars of as-cast and treated $Zr_{54.46}Al_{9.9}Ni_{4.95}Cu_{29.7}Pd_{0.99}$ BMGs tested in dry-sliding, respectively. (**b**,**d**) High-magnification detailed micrographs corresponding to the part marked by the red-lined rectangular in (**a**,**c**), respectively.

Figure 7a,c display the worn surfaces of as-cast and treated BMG in 3.5% NaCl solution, respectively, at low magnification. Owing to the lubricating effect of NaCl solution [30], these worn surfaces are covered with several narrow and shallow grooves and a few flake-like wear debris, which appear relatively smoother than those under dry sliding conditions, indicating slight abrasive and adhesive wear. Meanwhile, in order to observe the differences in the worn surface morphology of as-cast and treated BMGs more clearly, Figure 7b,d show detailed micrographs corresponding to the red-lined rectangular in Figure 7a,c respectively. Apparently, the wear debris on the worn surfaces of the treated sample are very similar to those of the as-cast sample.

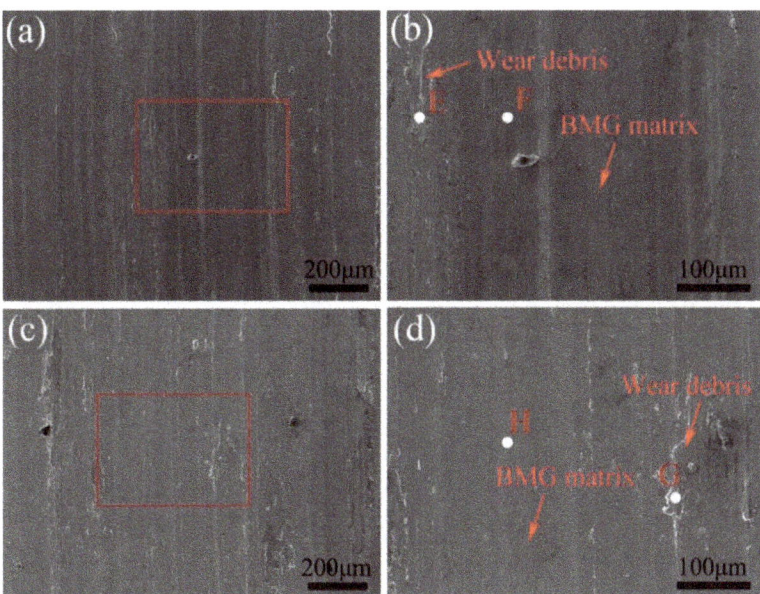

Figure 7. (**a**,**c**) SEM images of worn scars of as-cast and treated BMGs tested in 3.5% NaCl solution, respectively. (**b**,**d**) High-magnification detailed micrographs corresponding to the part marked by the red-lined rectangular in (**a**,**c**), respectively.

In addition to the worn surface morphology, the residual elements on the worn surface, which come either from the surrounding environment or the counterpart, are also an important basis for understanding the friction process [31]. Therefore, a SEM-EDS experiment was carried out. Figure 8 shows the SEM-EDS point spectrum of wear debris regions (marked by A in Figure 6b and C in Figure 6d and BMG matrix regions (marked by B in Figure 6b and D in Figure 6d). The Figure 8 shows that the oxygen appears on the worn surfaces of all samples, which confirms our previous supposition that surface oxidation takes place during the friction process [32]. Further observing BMG matrix regions, the oxygen content of the as-cast sample is 11.4%, while the treated sample is just 7%, suggesting that harder regions make it more difficult for the treated samples to participate in the oxidation reaction. Whereas as shown in wear debris areas, it is interesting to note that both as-cast and treated samples exhibit a larger amount of oxygen (over 50 at%) and the iron element, indicating that the wear debris would not only react violently with oxygen in the air, but also cause material transfer between BMGs and friction pairs. Furthermore, by comparing the variation in the chemical compositions of region A and region C, the oxygen content of wear debris increased from 52.3% in the as-cast BMGs to 61.3% of the treated sample, and the Fe content increased from 12.5% in the as-cast state up to 15.2%, contrary to the tendency of BMG matrix regions. This may be explained by the wear debris materials from the soft regions of worn surfaces; greater softness of the soft regions for treated BMGs results in their wear debris being more inclined to actively participate in the reaction.

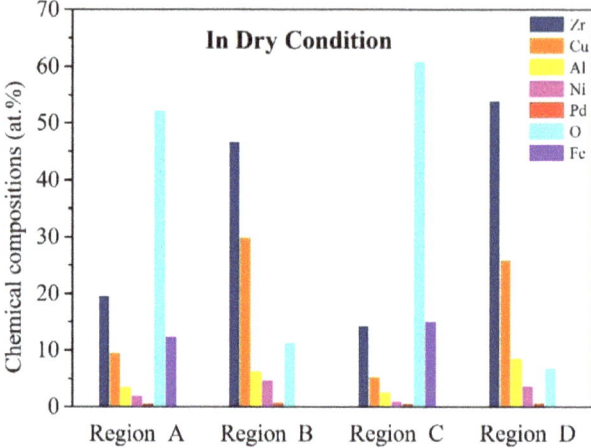

Figure 8. Chemical compositions of typical regions on the worn scar of as-cast and treated BMGs in dry-sliding.

As illustrated in Figure 9, the EDS analysis was performed under 3.5% NaCl solution on wear debris regions (marked by E in Figure 7b and G in Figure 7d) and BMG matrix regions (marked by F in Figure 7b and H in Figure 7d). It is obvious that the extremely low content of oxygen on the as-cast and treated BMG matrix regions is less than 2%, showing that the NaCl solution can reduce friction heat effectively, which minimizes the probability of an oxidation reaction on the friction surface [31]. Moreover, the chlorine element was not detected in this area, indicating that the corrosive effect of chloride ions could not have a significant impact on this area. Conversely, when observing the point scanning results of the wear debris regions, it is worthwhile to note the higher oxygen content and the appearance of chlorine elements. This implies that the wear debris is more easily oxidized and corroded as well, under the 3.5% NaCl solution. Furthermore, the contents of O and Cl on the wear debris of as-cast samples are both lower than the treated samples.

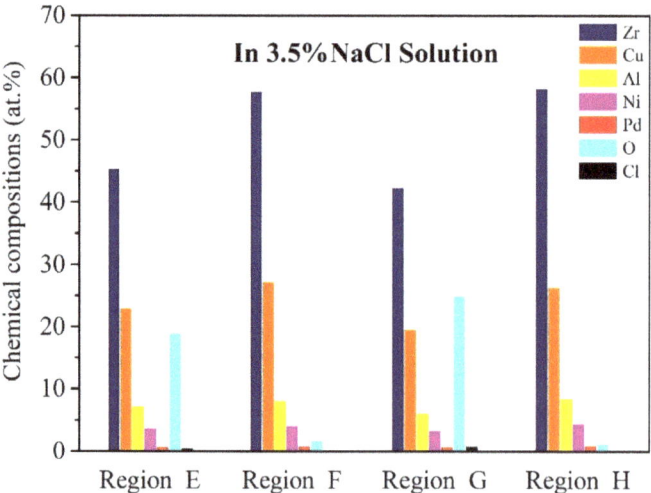

Figure 9. Chemical compositions of typical regions on the worn scar of as-cast and treated BMG in 3.5% NaCl solution respectively.

4. Conclusions

In summary, our observations revealed that wear resistance behaviors under the dry and the 3.5% NaCl solution conditions can be improved by imparting high structural heterogeneity to $Zr_{54.46}Al_{9.9}Ni_{4.95}Cu_{29.7}Pd_{0.99}$ BMG using the HRRF method. Under the dry condition, it is obvious that BMGs with different levels of structural heterogeneity showed significantly different wear resistance behaviors. The treated BMG exhibited a wear rate of ~14.7 × 10^{-6} mm^3·N^{-1}m^{-1} which is superior to the ~22.1 × 10^{-6} mm^3·N^{-1}m^{-1} wear rate of the as-cast BMG. Moreover, the predominant wear mechanism of all BMGs under dry-sliding conditions is characterized as abrasive and oxidation wear. Additionally, under the 3.5% NaCl solution condition, the main wear mechanisms are abrasive wear, slight corrosion wear, and adhesive wear. Due to the lubrication and cooling from the solution, although the wear rate of BMGs exhibits much lower levels than those of the dry-friction condition, the wear rate of treated BMGs (~3 × 10^{-6} mm^3·N^{-1}m^{-1}) was also below the as-cast samples (~4.5 × 10^{-6} mm^3·N^{-1}m^{-1}). Hence, the combination of improved wear resistance of BMGs both in dry and 3.5% NaCl solution conditions can be achieved. On one hand, the number of oxide layers generated on the worn surface of the treated sample were less than those of the as-cast sample. On the other hand, the higher inhomogeneous structure possesses higher ductility. It can effectively avoid the appearance of delamination, peeling, and cracks on the worn surface. This research provides an unprecedented perspective on BMGs, with easy implementation and celerity to improve wear resistance. We believe our approach presents a guide for the future development of BMGs.

Author Contributions: Data curation, Y.M.; Formal analysis, Y.M. and F.Z.; Investigation, Y.M.; Methodology, Y.M.; Resources, F.Z.; Writing—original draft, Y.M. and M.L.; Writing—review & editing, Y.M. All authors have read and agreed to the published version of the manuscript.

Funding: This research received no external funding.

Acknowledgments: The authors thank Xianping Wang and Yan Jiang at Chinese Academy of Sciences for the Nanoindentation test.

Conflicts of Interest: The authors declare no conflict of interest.

References

1. Jafary-Zadeh, M.; Kumar, G.P.; Branicio, P.S.; Seifi, M.; Lewandowski, J.J.; Cui, F. Critical review on metallic glasses as structural materials for cardiovascular stent applications. *J. Funct. Biomater.* **2018**, *9*, 19. [CrossRef] [PubMed]
2. Qiao, J.C.; Wang, Q.; Pelletier, J.M.; Kato, H.; Casalini, R.; Crespo, D.; Pinedag, E.; Yao, Y.; Yang, Y. Structural heterogeneities and mechanical behavior of amorphous alloys. *Prog. Mater. Sci.* **2019**, *104*, 250–329. [CrossRef]
3. Guo, S.F.; Zhang, H.J.; Liu, Z.; Chen, W.; Xie, S.F. Corrosion resistances of amorphous and crystalline Zr-based alloys in simulated seawater. *Electrochem. Commun.* **2012**, *24*, 39–42. [CrossRef]
4. Hong, W.; Baker, I.; Yong, L.; Wu, X.; Munroe, P.R. Effects of environment on the sliding tribological behaviors of Zr-based bulk metallic glass. *Intermetallics* **2012**, *25*, 115–125.
5. Hua, N.; Zheng, Z.; Fang, H.; Ye, X.; Lin, C.; Li, G.; Wang, W.; Chen, W.; Zhang, T. Dry and lubricated tribological behavior of a Ni- and Cu-free Zr-based bulk metallic glass. *J. Non Cryst. Solids* **2015**, *426*, 63–71. [CrossRef]
6. Khun, N.W.; Yu, H.; Chong, Z.Z.; Tian, P.; Tian, Y.; Tor, S.B.; Liu, E. Mechanical and tribological properties of Zr-based bulk metallic glass for sports applications. *Mater. Des.* **2016**, *92*, 667–673. [CrossRef]
7. Tian, P.; Khun, N.W.; Tor, S.B.; Liu, E.; Tian, Y. Tribological behavior of Zr-based bulk metallic glass sliding against polymer, ceramic, and metal materials. *Intermetallics* **2015**, *61*, 1–8. [CrossRef]
8. Segu, D.Z.; Hwang, P.; Gachot, C. A comparative study of the friction and wear performance of Fe-based bulk metallic glass under different conditions. *Ind. Lubr. Tribol.* **2017**, *69*, 919–924. [CrossRef]
9. Ishida, M.; Takeda, H.; Nishiyama, N.; Kita, K.; Shimizu, Y.; Saotome, Y.; Inoue, A. Wear resistivity of super-precision microgear made of Ni-based metallic glass. *Mater. Sci. Eng. A* **2007**, *449*, 149–154. [CrossRef]
10. Prakash, B. Abrasive wear behaviour of Fe, Co and Ni based metallic glasses. *Wear* **2005**, *258*, 217–224. [CrossRef]
11. Blau, P.J. Friction and wear of a Zr-based amorphous metal alloy under dry and lubricated conditions. *Wear* **2001**, *250*, 431–434. [CrossRef]
12. Louzguine-Luzgin, D.V.; Ito, M.; Ketov, S.V.; Trifonov, A.S.; Jiang, J.; Chen, C.L.; Nakajima, K. Exceptionally high nanoscale wear resistance of a Cu47Zr45Al8 metallic glass with native and artificially grown oxide. *Intermetallics* **2018**, *93*, 312–317. [CrossRef]

13. Ji, X.; Hu, B.; Li, Y.; Wang, S. Sliding tribocorrosion behavior of bulk metallic glass against bearing steel in 3.5% NaCl solution. *Tribol. Int.* **2015**, *91*, 214–220. [CrossRef]
14. Ying, L.; Pang, S.; Wei, Y.; Hua, N.; Liaw, P.K.; Zhang, T. Tribological behaviors of a Ni-free Ti-based bulk metallic glass in air and a simulated physiological environment. *J. Alloys Compd.* **2018**, *766*, 1030–1036.
15. Hua, N.; Chen, W.; Wang, Q.; Guo, Q.; Huang, Y. Tribocorrosion behaviors of a biodegradable Mg65Zn30Ca5 bulk metallic glass for potential biomedical implant applications. *J. Alloys Compd.* **2018**, *745*, 111–120. [CrossRef]
16. Rahaman, M.L.; Zhang, L. Size effect on friction and wear mechanisms of bulk metallic glass. *Wear* **2017**, *376*, 1522–1527. [CrossRef]
17. Ma, Y.B.; Jiang, Y.; Ding, H.P.; Zhang, Q.D.; Li, X.Y.; Zu, F.Q. Torsional behaviors in Zr-based bulk metallic glass with high stored energy structure. *Mater. Sci. Eng. A* **2019**, *751*, 128–132. [CrossRef]
18. Liu, Y.H.; Wang, G.; Wang, R.J.; Zhao, D.Q.; Pan, M.X.; Wang, W.H. Super plastic bulk metallic glasses at room temperature. *Science* **2007**, *315*, 1385–1388. [CrossRef]
19. Sopu, D.; Foroughi, A.; Stoica, M.; Eckert, J. Brittle-to-Ductile transition in metallic glass nanowires. *Nano Lett.* **2016**, *16*, 4467–4471. [CrossRef]
20. Wang, C.; Yang, Z.Z.; Ma, T.; Sun, Y.T.; Yin, Y.Y.; Gong, Y.; Gu, L.; Wen1, P.; Zhu, P.W.; Long, Y.W.; et al. A High stored energy of metallic glasses induced by high pressure. *Appl. Phys. Lett.* **2017**, *110*, 111901. [CrossRef]
21. Rahaman, M.L.; Zhang, L.; Liu, M.; Liu, W. Surface roughness effect on the friction and wear of bulk metallic glasses. *Wear* **2015**, *332*, 1231–1237. [CrossRef]
22. Huang, Y.; Fan, H.; Wang, D.; Sun, Y.; Liu, F.; Shen, J.; Sun, J.; Mi, J. The effect of cooling rate on the wear performance of a ZrCuAlAg bulk metallic glass. *Mater. Des.* **2014**, *58*, 284–289. [CrossRef]
23. Yang, H.; Liu, Y.; Zhang, T.; Wang, H.; Tang, B.; Qiao, J. Dry sliding tribological properties of a dendrite-reinforced Zr-based bulk metallic glass matrix composite. *J. Mater. Sci. Technol.* **2014**, *30*, 576–583. [CrossRef]
24. Parlar, Z.; Bakkal, M.; Shih, A.J. Sliding tribological characteristics of Zr-based bulk metallic glass. *Intermetallics* **2008**, *16*, 34–41. [CrossRef]
25. Jiang, X.; Song, J.; Fan, H.; Su, Y.; Zhang, Y.; Hu, L. Sliding friction and wear mechanisms of Cu36Zr48Ag8Al8 bulk metallic glass under different sliding conditions: Dry sliding, deionized water, and NaOH corrosive solutions. *Tribol. Int.* **2020**, *146*, 106211. [CrossRef]
26. Zhou, Z.; Wang, L.; He, D.Y.; Wang, F.C.; Liu, Y.B. Microstructure and wear resistance of Fe-based amorphous metallic coatings prepared by HVOF thermal spraying. *J. Therm. Spray Technol.* **2010**, *19*, 1287–1293. [CrossRef]
27. Zhang, T.; Lin, X.M.; Yang, H.J.; Liu, Y.; Wang, Y.S.; Qiao, J.W. Tribological properties of a dendrite-reinforced Ti-based metallic glass matrix composite under different conditions. *J. Iron Steel Res. Int.* **2016**, *23*, 57–63. [CrossRef]
28. Hua, N.; Chen, W.; Wang, W.; Lu, H.; Ye, X.; Li, G.; Chen, L.; Huang, X. Tribological behavior of a Ni-free Zr-based bulk metallic glass with potential for biomedical applications. *Mater. Sci. Eng. C Mater. Biol. Appl.* **2016**, *66*, 268–277. [CrossRef]
29. Lin, X.; Bai, Z.; Liu, Y.; Tang, B.; Yang, H. Sliding tribological characteristics of in-situ dendrite-reinforced Zr-based metallic glass matrix composites in the acid rain. *J. Alloys Compd.* **2016**, *686*, 866–873. [CrossRef]
30. Bakkal, M. Sliding tribological characteristics of Zr-based bulk metallic glass under lubricated conditions. *Intermetallics* **2010**, *18*, 1251–1253. [CrossRef]
31. Zenebe, D.; Yi, S.; Kim, S.S. Sliding friction and wear behavior of Fe-based bulk metallic glass in 3.5% NaCl solution. *J. Mater. Sci.* **2012**, *47*, 1446–1451. [CrossRef]
32. Ji, X.; Jin, J.; Tian, F.; Zhao, J.; Zhang, Y.; Yan, C.; Fu, L. Effect of cyclic cryogenic treatment on tribological performance of Fe-based amorphous coatings in air and in 3.5% NaCl solution. *J. Non Cryst. Solids* **2022**, *583*, 121471. [CrossRef]

MDPI
St. Alban-Anlage 66
4052 Basel
Switzerland
Tel. +41 61 683 77 34
Fax +41 61 302 89 18
www.mdpi.com

Materials Editorial Office
E-mail: materials@mdpi.com
www.mdpi.com/journal/materials

www.ingramcontent.com/pod-product-compliance
Lightning Source LLC
LaVergne TN
LVHW070609100526
838202LV00012B/605